AF598629

NAPIER UNIVERSITY LIS
3 8042 00555 6384

The CORPORATE HEALTHCARE REVOLUTION

Strategies for preventive medicine at work

David Ashton

Published in association with the
Institute of Personnel Management

The masculine pronoun has been used throughout this book. This stems from a desire to avoid ugly and cumbersome language, and no discrimination, prejudice or bias is intended.

First published in 1989 by
Kogan Page Ltd, 120 Pentonville Road, London N1 9JN
in association with the Institute of Personnel Management,
IPM House, 35 Camp Road, London SW19 4UW.

Printed and bound in Great Britain by Biddles Ltd
Typeset by the Castlefield Press Ltd, Wellingborough, Northants.

British Library Cataloguing in Publication Data
Ashton, David
The corporate healthcare revolution.
1. Great Britain. Industries. Personnel. Health
I. Title
613.6'2

ISBN 1-85091-694-2

To

Kerry, Bruce Owain and Luke

CONTENTS

LIST OF FIGURES

LIST OF TABLES

FOREWORD

For every working day lost through strikes, more than 30 days are lost through absence – at a cost of over £5 billion a year, equivalent to our national expenditure on non-military research and development. These simple facts in themselves mean that tackling absence should be high on the agenda of every employer.

Of course, some of that absence is voluntary, with poorly motivated employees simply opting not to turn up for work. But most of the working time lost through absence is a result of genuine sickness absence, and no amount of incentive schemes or disciplinary arrangements can alter that. What can make a difference, as more and more companies in this country and overseas are discovering, is positive action to promote good health among employees.

As David Ashton makes clear in this book, such programmes are not about firms acting as nannies to their employees. Nor need they involve massive expenditure. What they are, rather, are part of sound business strategy, part of the effective management of a firm's human resources with the aim of minimising the costs which ill health among employees imposes on every business and the economy as a whole. Every business must decide for itself whether such programmes are appropriate and, if so, what form they should take. Different approaches suit different employers and different workforces. What is clear is that those companies which have taken such initiatives have seen very real benefits to their businesses as well as to their employees. And that is a winning combination which cannot be ignored.

John Banham
Director-General
Confederation of British Industry

PREFACE

In writing this book I have four major aims.

First, briefly to discuss the origins and scale of our current health problems, both from a medical and an economic point of view.

Secondly, to provide some basic but essential background information about the principal risk factors to our health and how these can interact to produce the modern Western diseases. It is not possible to understand the scope of the current challenge or the opportunities for effective action without some knowledge of the disease processes we are attempting to combat.

My third aim is to provide a brief overview of the United States experience during the past eight to ten years. Workplace 'wellness' programmes have flourished there during this period; clearly there are some important lessons for us to learn and some mistakes we can avoid. A brief description of some of the more important programmes in the United Kingdom is also included. A growing body of evidence suggests that such initiatives are not only reflected in workforce health, but may also produce very significant financial benefits. In an increasingly competitive environment this is obviously of some importance, and I have attempted to provide a critical evaluation of these claims in the light of the available evidence.

My fourth and final aim (perhaps the most important), is to provide some practical suggestions as to how companies may establish their own worksite health promotion programmes with specific reference to the major diseases already discussed. I must emphasise that it is not possible within the scope of this book to go into great detail about the implementation of these programmes. This is because companies come in many forms and it would be absurd to recommend a specific formula in the belief that it would apply to all of them. My intention is to provide some general indications, a general framework within which appropriate initiatives can be taken based upon a clear understanding of the fundamental concepts involved.

This book is aimed at a wide audience which includes senior executives, personnel directors and human resource managers, and I have therefore not assumed any prior medical knowledge. Nevertheless I believe that the book will be of equal interest to doctors, nurses and the various paramedical personnel who work within our industries and who recognise the importance of health promotion in the workplace. The specific background and needs of the reader will obviously dictate how the book is used, and this is my intention. Some may wish to read the text from cover to cover, while others may wish to use it more selectively as a reference source for statistical information or to provide an overview of a particular health topic. I am conscious that some of my readers may feel a certain trepidation at having to deal with medical terminology but I hope that as they become familiar with the text, they will find that such fears were unfounded.

The book is divided into three major sections. Part 1 (chapters 1–3), reviews the origins and scale of our current health problems. It also discusses the US and the British experience of health promotion programmes, from both an economic and a medical viewpoint. I believe that this part of the book will be of interest to all groups, but particularly to senior managers and executives who have to confront issues relating to the financial implications and potential economic benefits of such programmes.

Part 2 (chapters 4–8), provides some background information concerning five major risk factors which may interact to produce various diseases, but which can be favourably influenced by worksite health promotion activities. I believe that this part of the book will be of interest to all readers irrespective of whether their viewpoint is primarily corporate, medical or individual. Moreover, this background knowledge is fundamental to a clear understanding of the preventive strategies discussed in Part 3 of the book.

Part 3 (chapters 9–14) shows how the five major risk factors discussed in Part 2 interact to contribute to a number of major diseases such as cancer, heart disease and stroke. In particular, it discusses the practical application of some health promotion strategies designed to reduce the prevalence of these diseases within the workforce and the population as a whole. This will obviously be of particular interest to personnel directors, human resource managers, doctors, nurses and paramedical personnel who are likely to be closely involved in the planning, implementation and administration of such programmes.

It is perhaps appropriate at this point to say something about occupational medicine, which has long been regarded as the medicine of the workplace. The relationship between occupation and health has, of course, been recognised for centuries. Indeed the first occupational disease was probably silicosis (a lung disease), which occurred in the Stone Age communities which made flint tools and weapons. Nowadays we are much more familiar with occupational medicine as it applies to coal miners, workers in the nuclear industry and other occupations where environmental hazards are apparent. Health and safety at work, accidents and trauma, skin diseases, noise control, the handling of dangerous substances and other occupational hazards are the traditional province of what we understand as occupational medicine.

This book, then, is not about occupational medicine as it has traditionally been regarded. Rather it is about how we can influence those major determinants of health which are fundamentally related to personal habits and behaviour.

ACKNOWLEDGEMENTS

Writing a book involves the input of many individuals, most of whom receive little or no recognition for their efforts. I am acutely concious of standing on the shoulders of others; a vast army of scientists, physicians and researchers who have committed so much of their talent, energy and hard work to creating the enormous body of scientific evidence upon whose foundation much of this book rests. Without them none of this would be possible and I am pleased and privileged to acknowledge my indebtedness.

On a more personal note, there are several individuals who deserve special mention. I am very grateful to Bob Mathers and John Jackson for providing invaluable criticism and comments on the manuscript prior to the final draft. They raised a number of important issues, and were instrumental in shaping the final product. Thanks also to Stephen Collier for his advice on specific sections of the text.

I am also pleased to be able to thank some of the many professional colleagues who have contributed. They include Dr. Bruce Davies of AMI Healthcare plc; Dr. Derek Taylor of Marks and Spencer plc; Dr. Richard Welch of The Post Office; Dr. Nils Stormby of Medscand, Sweden; Dr. Audrey Tucker of St Bartholomew's Hospital; Dr. Pat Last of BUPA; Dr. Derek Rowlands of Manchester Royal Infirmary and Dr. Tony Rickards of the National Heart Hospital.

I am grateful to John Banham, Director General of the Confederation of British Industry for providing the Foreword to the book, and also for the support of the Institute of Personnel Management.

I owe a special debt to Pat Parker, who devoted so much of her own time and energy to the typing and preparation of the manuscript, that it would be difficult to find words adequate enough to express my thanks. Without her hard work, patience and attention to detail, the project would not have been completed. I want her to know how much I appreciate it.

Finally, thanks to my wife Kerry. Through our discussions together she contributed a great deal to this book, and her love, patience and encouragement supported me through many long hours.

INTRODUCTION

The ideas and arguments presented in this book are underpinned by two fundamental propositions:

1. That much of our current disease burden is preventable: what determines the health of human beings is not primarily due to medical intervention, but to what we eat, how we behave and our environment.
2. That the workplace is an extremely powerful delivery point for preventive programmes and health education.

The idea that our modern diseases of affluence may be in large part preventable is not a new one. Indeed, the importance of prevention is enshrined in the original Beveridge Report of 1942, which set the foundation for our current National Health Service. Given that the NHS will shortly be undergoing the most radical reform since its inception, it is perhaps an appropriate moment to ask whether, and to what degree, the Beveridge vision has been fulfilled. The report sets out the major objectives of the future medical service as being 'to provide a system of medical service directed towards the achievement of positive health, of the prevention of disease, and the relief of sickness' (Part VI, s. 427).

It is absolutely clear that for the past 40 years or so we have concentrated virtually all of our skills and resources on the last of these objectives, almost completely ignoring the other two. There is now an enormous and quite unjustifiable disparity between what we spend on the treatment of established disease compared with what we spend on trying to prevent the disease occurring in the first place. The most graphic illustration of this is that although the total spending on the National Health Service in 1989 will be approximately £26 billion, only £20 million will be allocated to the Health Education Authority. This represents only about .07 per cent of the total spending on health care. It is therefore more appropriate to regard our current system as a National *Disease* Service than a National Health Service.

More worrying still is the fact that the recent White Paper on the NHS contains hardly a passing reference to the need for prevention. This is despite the fact that in the 40 years since the publication of the Beveridge Report, an enormous amount of scientific data have been collected which powerfully support the case for prevention. That it has been so completely neglected in the current report is scandalous. It is clear that there will have to be a major shift in attitude and priorities among those who are currently responsible for health policy in this country.

My own change in attitude relates to an apparently minor incident which occurred a few years after I qualified while I was working as a medical registrar in a large hospital in the North of England. In those days it was quite common (it probably still is) for junior hospital doctors to carry out GP locum work in the evenings or weekends as a means of gaining some valuable additional experience (and also some valuable additional cash!). I arrived at a local practice for an evening surgery, and after a quick cup of coffee I pressed the buzzer for the first patient. She was a woman of about 30, well dressed and quite well spoken and after she had taken a seat I asked her how I could help. Without any hesitation, and smiling broadly, she told me she had come for her 'happy pills'. From her notes I gathered that she had been taking tranquillisers for some years, although it wasn't very clear why she had been prescribed them in the first place.

During the course of the next few minutes I tried to understand something of the patient's circumstances, and in particular why she felt the need to be taking medication so regularly and over such a long period. The minor disappointments and frustrations of everyday living were there, but certainly no more than most people have to contend with. I explained to her that these tranquillisers, mild though they may have seemed, had significant addictive qualities and that it would be much better for her to consider coming off them altogether. This proved an unpopular suggestion: she became quite aggressive and told me that I had no good reason for refusing to renew her regular prescription. She had been taking these tablets for a number of years without any apparent side-effects; besides which, she would feel unhappy without them. I tried to explain that while I understood her reaction, I did not feel that this in itself was a good enough reason to go on prescribing as previously. This produced an even more aggressive reaction. After calling me a few choice names, she slammed the door loudly on her way out.

At the time I didn't think a great deal of it: patients quite commonly insult their doctors and for some it may actually be part of their therapy! But on reflection later that evening, I recognised something of fundamental importance about the whole episode. It was a graphic example of the degree to which we have become 'medicalised'. So far as this lady was concerned, to be unhappy was to be abnormal.

I suppose it struck a particular chord with me at that time because I had recently read Ian Kennedy's book, *The Unmasking of Medicine*, published in 1981 and based on the Reith lectures which Professor Kennedy had delivered in 1980. One of the arguments in his book is that doctors have in some senses become agents of social control. Many patients who visit GP's surgeries are not suffering from any form of medical illness whatever. Their main problem is that they are unemployed, or have unsatisfactory housing, or are depressed because of difficulties with relationships and so on. These problems find their origins in the patients' economic and social circumstances rather than being primarily determined by specific medical causes. Ultimately, of course, these adverse social and economic factors are linked to the prevailing political environment. By doling out tranquillisers, Kennedy argues, the doctor is unwittingly helping to maintain the social, political and economic status quo. What I had experienced that day was just one example of the readiness with which individuals will allow, even demand, that the doctor should exert such control.

I found this disturbing because it seemed to me a powerful example of the extraordinary degree to which we have become dependent upon our medical system – and upon doctors in particular. Why this dependence should have occurred is complex and goes to the very heart of the way in which we as a culture think about our bodies, health and disease. But there is no doubt that it is in part created by the apparent success of modern medical science.

One of the consequences of the remarkable progress made by the NHS has been the abdication of the individual's responsibility for his own health. Because the state became responsible for the health of the population, the state was also responsible for *each person's* body: it was up to the state to make sure that it was put right if anything went wrong. And not only is physical wellbeing the responsibility of the state, so too is individual happiness or unhappiness. In practical terms, this attitude translated itself into a belief that an individual's physical and mental health could be left in the hands of the doctor. As we shall see later, it is an unrealistic and unreasonable expectation; it is also one which, it must be said, doctors have been instrumental in creating.

However unpleasant it may be, the fact is that there is a very poor relationship between our current system of healthcare and the health of the population as a whole. More medicine always means greater cost, but it does not necessarily mean better health. On the contrary, our health is primarily determined not by medical intervention, but by what we as individuals do to ourselves. The overemphasis on the need to treat established disease, whether by surgery, drugs, radiation or some other form of therapy, has led to a dangerous tendency to ignore the very factors which made us ill in the first place.

The major causes of premature death and disability in our society, and in Western society in general, are the so-called diseases of affluence. We are only too familiar with them. Cancer, heart disease, strokes, mental illness, obesity and the like are examples of diseases which may be more insidious, but are no less devastating than the infectious diseases they now replace. There is now compelling evidence to suggest that much of this disease burden is preventable and that what is needed is a nationally coordinated, properly resourced programme of prevention – one that will reach all sections of the community, including the very young and the old. In committing so many of our resources to technological medicine and what is essentially a sickness service, without proper attention to the need for prevention, doctors have committed a grave error.

It is worth reminding ourselves that the word 'doctor' derives from the Latin word *docere* which means 'to teach'. But how much do our doctors teach us about health and how to avoid disease? We all know what the answer is – especially as doctors themselves are not always the best examples of a healthy lifestyle. Part of the problem here is that medical education itself is so powerfully orientated towards diagnosis and treatment that prevention has been relatively neglected. In my own training, for example, there was scarcely any reference whatever to the prevention of disease throughout six years at medical school. Given the weight of evidence supporting the case for prevention, this is astonishing. Not only do we need a fundamental rethink of the way in which medicine is practised in this country, there also needs to be much more emphasis placed on prevention in the curriculum of our medical schools.

My own view is based on a simple proposition: it is infinitely better to prevent a disease than to commit massive resources in an attempt to cure the damage once it has occurred. With few exceptions, the word 'cure' is actually inappropriate. Many, perhaps most, minor diseases are entirely self-limiting and will correct themselves without any medical intervention. In the case of more serious diseases the damage has often been done and the disease well established by the time the doctor is consulted – thus making a cure, in the real sense of the word, difficult or impossible. In no way does this diminish the importance of the doctor's role in relieving distress and prolonging life. But this is not my argument. The point is that to focus exclusively upon the 'curative' part of medicine is inappropriate and unhealthy.

The need for a nationwide initiative in preventive medicine has never been more urgent. For a whole variety of complex reasons which will be discussed later, the NHS is unable to provide such an initiative and the government is unlikely to provide any additional resources to make it possible. It might be argued that general practitioners should be responsible for this and there is no doubt that they have a major part to play. But the problem is that most general practitioners are so heavily committed to the management of the sick that they have neither the resources nor the time to devote to the promotion of health or the prevention of disease.

The central thesis of this book is quite simple. I believe that corporate organisations have a unique opportunity to exert a major – perhaps even a decisive – effect on the health of the population in the 1990s and beyond. A 'Corporate NHS', focused on the prevention of disease and the promotion of health in the workplace, is not only medically desirable but also commercially and economically sensible. Indeed, the incentive for companies and organisations to act has never been greater. The Confederation of British Industry (CBI) estimates that absenteeism costs this country in excess of £5 billion per year. In fact, for each of the 3.5 million working days lost through strikes in 1987, more than 30 were lost due to absence. Moreover, according to available data, the UK appears to have a relatively high absenteeism rate compared to its major trading competitors. The results are fairly predictable: lower output, poorer service, lost orders and, as a result, fewer jobs. While there is no single strategy for improving attendance, there is no doubt that measures designed to prevent disease and promote health in the workplace can exert a major impact in reducing absenteeism rates due to ill-health. It follows that responsible employers cannot but be interested in the health of their employees: it not only makes good medical sense – a failure to do so may have serious financial consequences.

Attitudes towards health are already beginning to change, and many individuals and organisations have recognised the importance and the value of encouraging healthy lifestyle practices. This is clearly an encouraging trend and once which will doubtless continue and grow. What is not so encouraging, however, is that these shifts in attitude and behaviour are not apparent across all social groups. They are most evident in the middle class, relatively affluent and well educated members of our society. A similar social class gradient has also been observed in other countries where changes in behaviour and attitudes towards health issues have occurred.

There are two possible explanations for this uneven social distribution. Either people in the lower social groups with relatively poor educational standards are simply not interested in health, or else the health education message was written *by* the affluent and well educated *for* the affluent and well educated and is therefore replete with all the social, economic and political values which that implies. It is of the greatest importance that we develop a *democracy* of prevention, one accessible to and understandable by everyone, including the poor, the unemployed, the elderly and the less educationally advantaged.

The choice which faces us now could not be more stark. We either continue to support our present sickness service with its overemphasis on medical technology or we move towards a more balanced, holistic and humane approach which recognises the importance of both treatment and *prevention* in equal measure. The decision is an important one because in choosing for ourselves, we are also choosing for our children and future generations. My message is expressed succinctly by Stafford Beer in his book *Platform for Change*: 'We must invent the future. If we do not the future will happen to us. We shall not like it.'

PART 1

CHAPTER 1

CRISIS AND OPPORTUNITY

'Let the great world spin forever down the ringing grooves of change'
Tennyson

If one had to select two words to characterise the pre-occupations of most corporate organisations during the 1980s, they would almost certainly be 'change' and 'innovation'. Change, in particular, is the buzz-word of the 1980s. Indeed it is scarcely possible nowadays to pick up a business journal or management handbook without reading about the effects of change in one form or another. We talk incessantly about the management of change, we have lectures, workshops and seminars on how companies and individuals can better equip themselves for change. So much so that the word has become almost a cliché in itself.

To use another cliché: change has always been with us. Our bodies are in constant state of flux and evolution and we all experience change on a day-to-day and even a moment-to-moment basis. Corporate organisations are not so much things as processes, with mergers and takeovers occurring as normal events in the evolution of many companies. There is thus a natural cycle of growth and development, decay and subsequent rebirth. This is true not only of corporations and individuals but of whole civilisations. In this sense, therefore, our own particular period in history is no different from that of any other.

But what does mark it out as profoundly different is *the rate of change*: the sheer pace of life nowadays, the extraordinary rapidity with which even the most familiar aspects of our everyday lives appear to change almost overnight. Both the rate of change and the growing complexity of the innovations which result from it, place greater and greater demands upon individuals and organisations. They need the flexibility to adapt to the new challenges, threats and opportunities which change inevitably brings.

Nowhere is the phenomenon of change better illustrated than in the field of medical science. In many areas modern medicine has achieved a degree of success which is nothing less than spectacular. Developments in anaesthesia now permit surgical procedures of astonishing complexity. Heart, kidney, liver and lung transplants are now commonplace and in years to come we shall probably see organ banks being set up, with the tissues of the various organs already classified and awaiting an appropriately matched recipient.

Advances in genetic engineering have made it possible for scientists to manipulate the very substance of life itself; indeed the 'designer' gene is not far away. Currently, a number of research teams in various parts of the world are working towards providing a complete genetic map (genome): a 'blueprint' of a complete human individual. This will allow scientists selectively to alter and engineer specific parts of our genetic make-up; the potential consequences are difficult to imagine. The discovery of X-rays was a major milestone in medical history but today's 'imaging' technologies, with their space-age names, are in an altogether different league. Today it is computerised axial tomography (CAT), and tomorrow it will be magnetic resonance imaging (MRI) and positron emission tomography (PET). These modalities will, in turn, be superseded by even more esoteric technology.

The pharmacuetical industry, a phenomenon of the past 40 years or so, provides the medical profession with a truly bewildering variety of drugs and medicines which have revolutionised the management of many serious diseases. A whole new generation of biologically engineered therapeutic agents is beginning to emerge, with the advantage of being more specific in action while having far fewer side-effects than those drugs currently available.

The developments in surgery, genetics, drugs and other forms of medical technology have added enormously to our growing armamentarium against disease, and we are no longer surprised when we hear about the latest medical 'breakthrough'. There is an almost unshakeable belief on the part of the public that medical science will always find an answer and that eventually we shall have a cure for everything. It is often thought to be merely a question of money and time before we usher in the medical millennium and enjoy a long, happy and essentially disease-free existence.

That is what we would like to believe. Unfortunately, the reality is rather different.

Crisis

Despite all the advances in medical science there appears to be a growing and profoundly disturbing disparity between the development of high-technology medicine and the health of the nation. Indeed, given the current roster of diseases, it is difficult to understand how we might even begin to describe ourselves as healthy.

At the beginning of the century, by far the main causes of premature death were the infectious diseases, particularly typhoid, diphtheria and dysentery. These names would have struck fear into almost anyone at that time and yet nowadays we scarcely give them a second thought. It is often claimed that major developments in medicine – and in particular the introduction of antibiotics – were instrumental in bringing about the demise of these infectious diseases as a major cause of death. But this is not the case. The most important factors were actually social and economic: major improvements in housing, better standards of nutrition and the advent of a public health system, including more efficient sewage disposal. As indicated in Figure 1.1, the substantial reduction in death rates which was a feature of the nineteenth century occurred well before the introduction of antibiotics in the 1940s. Whatever the reasons, in Western industrialised societies these diseases are no longer a major cause of illness and death.

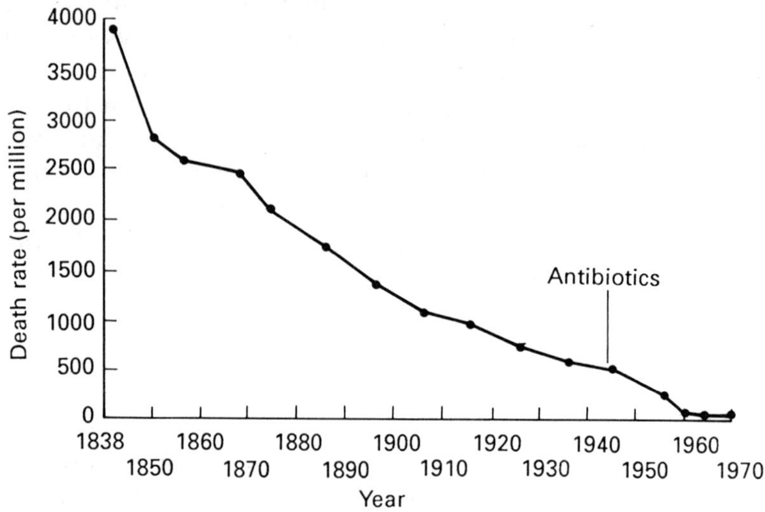

With permission of Blackwell Scientific Publications

Fig. 1.1 *Deaths from tuberculosis and the introduction of antibiotics*

Unfortunately we have replaced the infectious diseases with a variety of newer diseases such as cancer, heart disease, strokes, obesity, alcoholism and other forms of drug dependence which are more insidious but no less damaging or destructive. They are sometimes euphemistically referred to as the diseases of 'civilisation' or 'affluence'. As we shall see later, they are actually better regarded as diseases of *maladaptation* because they often arise from our inability to adjust biologically to rapid changes in our environment and living habits. One must, therefore, ask the question why, given the enormous sophistication of modern medicine and the whole panoply of hospitals, doctors, nurses, drugs and so on at our disposal, we should still be afflicted with this terrible and growing burden of disease? Part of the answer must be that in many ways our health system has failed us.

There is a perception on the part of many people, including politicians and some healthcare professionals, that the problem is merely one of inadequate funding: if the appropriate investment were made the problems of the NHS would be largely resolved. Others believe that increased funding should be combined with a streamlining process that would result in more doctors and nurses and less bureaucracy in the form of administrators and managers. Others, myself included, are not quite so optimistic.

The plain fact is that more doctors and nurses will not necessarily mean better health. What is wrong with our current system of healthcare is much more deep-seated than merely the question of government funding, administrative structure or the shortage of doctors and nurses. It goes, in fact, to the very roots of the way in which we as a culture think about health and disease. We need, I think, to ask ourselves how we reached this point and how we came to depend so heavily upon a system which appears, in the final analysis, to be failing us in so many ways.

The National Health Service

Throughout the Second World War plans were being drawn up for the creation of a 'new' Britain. The government had appointed a committee to undertake a comprehensive survey of existing national schemes of social insurance and allied services. The results of this committee's deliberations were subsequently published in the Beveridge Report of 1942. A document of enormous historic importance, it attempted for the first time to deal comprehensively with the problem of poverty. The main grounds for criticism of the pre-war medical services were that many patients were inadequately treated or were placed in financial difficulty by the cost of treatment. In 1938 the benefits provided under National Health Insurance were available to about 20 million insured workers, but rather more than half the population had to meet their own costs or seek assistance from the poor law, a system of relief which dated from Elizabethan times (1601).

The Beveridge Report was principally concerned with the provision of a minimum income for all through an effective system of social security. However, social security was to be only one component of a much wider plan which would also include economic policies designed to ensure the maintenance of full employment, family allowance and, of course, a National Health Service. The most important reason for introducing a comprehensive NHS was the fact that a large number of people found great difficulty in paying for medical care.

The vision of Aneurin Bevan was realised with the setting up of the NHS in 1948. There can be no question that the creation of the NHS provided a much improved system of medical care for the British population and it was women and children in particular who gained from its introduction. There would be little argument either that its most tangible achievement was to make medical services available to everyone and to remove the burden of direct payment from the large number of people who simply could not afford it.

The years immediately after the war were years of uncertainty, with a good deal of unemployment and no welfare state to provide appropriate support. After the Beveridge Report all this changed and people began to believe that they could expect very rapid and dramatic improvements in social conditions and employment. When the NHS was established there was little perception of what it could do; certainly no dependency culture was evident. However, as the service began to take up the slack, a notion developed that the state, via the NHS, was primarily responsible for an individual's health. The idea of a 'cradle to grave' system of care became widespread and has since grown dramatically. The perception that the NHS could deliver health to the whole nation was further reinforced by the seemingly invincible march of technology. The infectious diseases had all been overcome and quite astonishing advances were taking place in medicine and medical technology. This was only one component of a much broader technological revolution which appeared to reach its zenith with the Harold Wilson government of the 1960s and the 'white heat of technological revolution'.

Developments in medical technology were high profile. Because of growing public expectation it became politically advantageous to be seen to back these developments and, by inference, to have a positive impact upon the health of the population. The sheer magnitude of the success of modern medicine and the NHS seemed to make adequate financing a moral imperative that no government could ignore. The dependency of the general public on the NHS and the belief that health was the responsibility of the doctor – and eventually the state – became deeper and more widespread.

For some individuals it was not merely the state's responsibility to keep them healthy, it was also the state's fault when they became sick. This curious attitude declares that my health is anyone's responsibility *except* my own. It is exemplified by the fact that a number of individuals have recently attempted to sue tobacco companies for compensation for diseases they claim to have developed as a direct result of cigarette smoking. The same sort of logic would allow us to take food companies to court for selling unhealthy food which may contribute to heart disease, or motor manufacturers to court for providing us with the means to kill many people on our roads each year. Even accepting that the advertising industry does have some influence on our behaviour, the logic of this attitude is difficult to understand. What is missing from from all this is *the individual's responsibility for his own actions.*

This is not to say that the state does not have any responsibility; clearly it does. The responsibility of government is to make available to individuals the information necessary to make informed choices. This is clearly expressed in the DHSS consultative document,

Prevention and Health: Everybody's Business (1976). As the authors state:

> the prime responsibility for his own health falls on the individual. The role of the health professionals and of government is limited to ensuring that the public have access to such knowledge as is available about the importance of personal habits on health and at the very least, no obstacles are placed in the way of those who decide to act on that knowledge.

The biomedical model

One of the crucial factors in determining our view of the body and of disease is now referred to as the 'biomedical' model. In essence, this views the body as a highly complex machine, made up of a very large number of component parts. By studying each of these component parts in ever greater detail, medical science can achieve a deeper understanding of what makes the body function normally, and therefore how to correct these various 'mechanisms' when something goes wrong. In this sense, the perceived role of the doctor as some form of mechanic, is not far-fetched.

The biomedical model of disease has its origins in the work of the seventeenth-century French mathematician and philosopher René Descartes. There are two aspects of Cartesian thought which are of particular significance in our own discussion. First, Descartes' method was analytical and reductionist. It consisted of breaking up thoughts and problems into smaller pieces and arranging them in a logical order. As a result of this, his view of the world was intensely mechanistic, a view which he also extended to include plants, animals and human beings:

> I consider the human body as a machine . . . my thought . . . compares a sick man and an ill-made clock with my idea of a healthy man and a well-made clock.

And again:

> We see clocks, artificial fountains, mills and other similar machines which, though merely man made, have nonetheless

> the power to move by themselves in several different ways . . . I do not recognise any difference between the machines made by craftsmen and the various bodies that nature alone composes.

The second important contribution Descartes made was the view of the body as being essentially separate from the mind. In philosophical terms, the mind for Descartes was a more certain entity than the body and he considered them to be fundamentally different. 'There is nothing included in the concept of the body that belongs to the mind; and nothing in that of the mind that belongs to the body.' As we shall see later, the Cartesian division between mind and body has had a profound effect on our culture and in particular our view of the body and disease.

Descartes' great contribution to science was to provide the basis for analytical method and the general framework of a world view that regarded nature as a perfect machine governed by exact and noble mathematical laws. People, being but components of that world view, were governed by similar laws. It was left to the genius of Sir Isaac Newton to provide the detailed mathematical theory that would support Descartes' original framework. This intensely mechanistic and analytical view of the world has dominated our thinking in all the sciences for the past 300 years. As far as medicine is concerned, there are two important consequences.

The first is the extraordinary degree to which the idea of the body as a machine has become ingrained. This has certain inherent dangers, the most obvious of which is dehumanisation: the idea that people are *nothing but* machines. This gives rise to one of the most powerful and consistent criticisms of modern medicine – namely that it can alienate and dehumanise.

Secondly, because of the Cartesian division between mind and body there has been a tendency to concentrate mainly upon the physical aspects of disease, and to pay less attention to the psychological dimension. Yet it is a matter of common experience to us all that our emotional and psychological state exerts a powerful influence upon our sense of physical wellbeing. In fact, the link between the mind and various forms of physical disease has been observed for centuries; it is all the more surprising, then, that modern medicine has, until recently, paid relatively little attention to it. Nowadays we recognise that the mind and the body are not separate entities, but should be considered as a unity, a single complex system.

Investment in prevention

The development of modern scientific medicine based upon the biomedical model has, in the last two decades in particular, produced remarkable advances in our understanding of illness. These are of enormous importance and potential. Nowhere is this more apparent or dramatic than in the management of acute, life-threatening events such as major trauma resulting from road-traffic accidents, serious burns and other forms of acute medical catastrophe. In these circumstances the power of medical technology is breathtaking and it is no exaggeration to say that nowadays many people survive injuries that even five or ten years ago they would undoubtedly have died from. This is an area in which modern medical technology has triumphed and in which the service provided by the NHS has been a resounding success.

The problem is that the major causes of premature death and disability in our society are not due to acute life-threatening conditions but rather the chronic degenerative diseases of maladaptation. The application of sophisticated medical technology in these situations is much less dramatic and certainly cannot be regarded as curative.

That prevention is better than cure is self-evident – and nowhere more so than in the case of our modern-day disease. It is all the more relevant in the light of the large and growing body of evidence suggesting that prevention is a real possibility and that much of our current disease burden is, therefore, avoidable. We can only begin to make this a reality if we are prepared to commit sufficient resources to the task. However, there is no suggestion that these resources will be made available via our current healthcare system. Compared with the total amount of money allocated to the NHS, the amount which is actually spent on health education and prevention is frankly irrelevant. In the last 20 years or so, total spending on the NHS has risen more than twelve-fold, so that in 1988/89 it cost us no less than £26 billion. (See Figure 1.2.) In contrast, the amount spent on health

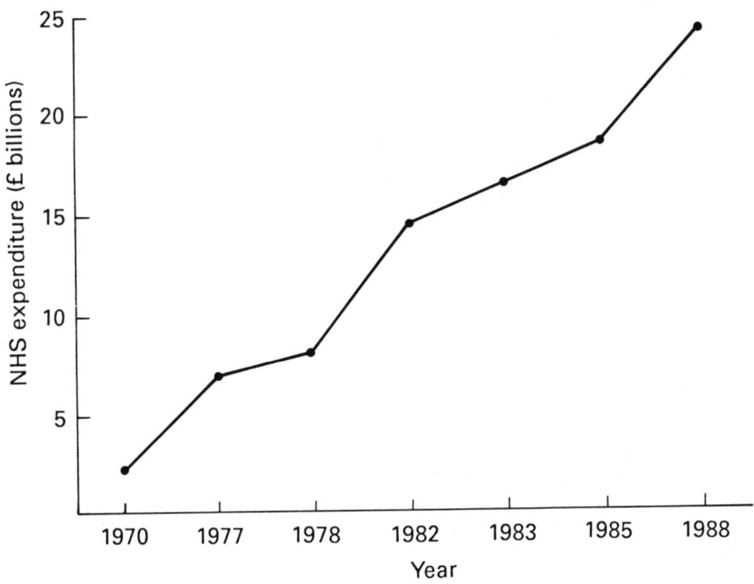

Fig. 1.2 *NHS spending: 1970–1988*

education during that same period has remained at *less than one-tenth* of 1 per cent of total NHS funding. (See Table 1.1.) This disparity is patently absurd. We could perhaps derive some comfort from the belief that our NHS is both popular

Table 1.1 *Funding of the Health Education Authority (HEA) as a percentage of total NHS funding*

Year	HEA allocation as percentage of NHS
1970	0.015
1977	0.022
1978	0.039
1982	0.045
1983	0.052
1985	0.052
1988	0.076

and effective in terms of producing a healthier population. Unfortunately, the evidence would suggest otherwise.

In 1983, the Annual British Social Attitudes Series asked the following questions: 'How satisfied are you with the way in which the NHS runs nowadays?' About 55 per cent of the population said they were satisfied, with only about 25 per cent registering as dissatisfied. By 1988, the situation had changed considerably, with only 37 per cent registering as satisfied and no less than 46 per cent as dissatisfied. Most significant of all is the number of individuals who said they were *very* dissatisfied: this had risen from 7 per cent in 1983 to no less than 25 per cent in 1988. The figures speak for themselves, revealing a serious level of concern on the part of the general public about the quality and direction of our current system of healthcare.

The Health and Lifestyle Survey published by the Health Promotion Research Trust in 1987, was equally revealing about the health of the population. Part of the survey included questions relating to the individual's perception of his/her own health status and participants were asked to report themselves as being in excellent, good, fair or only poor health. At first sight it may seem that such self-assessment measures are of doubtful validity but there is evidence that they correlate very well with health assessed in more objective ways. No less than 28 per cent of all males and

25 per cent of all females felt that they could rate their health as only fair or poor. There was also a very strong social-class gradient. For example, in males over the age of 50 in the lowest income group, 38 per cent reported their health as being only fair or poor, compared with 13 per cent in the highest income group. The survey also included some very interesting information with regard to the consumption of drugs. Thirty per cent of all males and 40 per cent of females are currently taking medication for one reason or another. These data do not suggest a high degree of success in our attempt to create a healthier nation, nor do they lend support to the frequent claim by some politicians that we are healthier now than ever before.

Politicians are also fond of claiming that we are living longer than at any other time in history and that this is a graphic expression of the success of the NHS. The facts are not quite so convincing. The truth is that although there has been a significant increase in life *expectancy* there has not been a significant increase in life *span*. In other words, many more people nowadays can expect to survive into old age (increased life expectancy), but those of us who do are not actually living all that much longer (life span). In mid-Victorian times, for example, a man of 50 could expect to go on living about another 20 years. Well over a century later, a 50-year-old man has, on average, 23 more years to live.

An extra two or three more years' life for a part of the population does not seem a great deal to show for the staggering increase in spending on the NHS. Nor should we assume that what increase in life span has occurred is due wholly to improvements in medical practice. Equally important have been improvements in social conditions, including housing and nutritional status. The life span of adults in the US has not improved much either, despite the fact that expenditure on health has risen from $12 billion in 1950 to over $275 billion in 1987.

The reason why life expectancy has increased is that fewer people nowadays are dying at birth or during infancy. In

Victorian times, for example, four out of ten children died in infancy. Today, that situation has changed dramatically, primarily due to improvements in hygiene, social circumstances, nutrition and improved medical care. As a result, the greatly reduced death rates among children and young people mean that more people can expect to survive into old age.

The 'corporate' NHS

Despite the urgent need for a major government initiative in the area of prevention, current attitudes indicate that this is likely to be a long time in coming, if it ever comes at all. Because our current health system is so strongly disease-orientated, there is by no means any guarantee that the medical profession would actually support it if it did happen. Some individual general practitioners do make extraordinary efforts to provide some basic programmes, usually with the assistance of the practice nurse, but unfortunately such initiatives are patchy in their distribution and there is very little coordination.

I believe that there is a major opportunity for private industry in this country to establish a network of preventive healthcare programmes that could well exert a decisive influence on the health of our population. Corporate organisations are fond of claiming that the most important asset they possess is their workforce. Indeed, the saying 'people are our most important asset' has become one of the modern-day corporate mantras. A few moments of reflection should be enough to convince anyone of its obvious truth. If one were to ask any employee in a company what his or her own most important asset was, they would almost certainly reply that it was their personal health. If personal health is important to all of us, and if the most important asset of any company is the people it employs, it must follow that all corporate organisations should be interested in the health of their employees. There are several more specific reasons in support of my argument.

First, I believe that companies should take the lead in developing a national initiative in disease prevention and health promotion, simply because no one else is going to do it. Preventive medicine in the workplace is not only a means of improving the health of the workforce, it can also be a powerful vehicle by which to enhance the company image and build bridges into the local community.

The second reason is that, in many respects, corporate organisations are in a unique position to influence the health of the workforce. Studies show clearly that in order to change behaviour one must continually reinforce the health education message over a long period. Corporations are in an ideal position to do this: on average, their workforce is present for eight hours a day. This captive audience means that there is an endless number of opportunities to reinforce the particular health-education message in a variety of ways. It is something which the general practitioner cannot do, simply because he could not reach that number of people – even if he had the inclination to do so. He certainly could not reach them on a regular enough basis.

More importantly, in order to bring about permanent changes in behaviour one must try to create an atmosphere and a culture which is supportive to the kind of behavioural change one is trying to bring about. For example, if I am a heavy cigarette smoker and I really want to give up, it is infinitely more difficult for me to do this if all my workmates continue to smoke while I am at work. The atmosphere is not conducive to my giving up and my chances of so doing are substantially reduced. If, on the other hand, the prevailing culture is supportive – if most of the people in my place of work are non-smokers – then my chances of giving up are substantially greater. Companies can and do create a specific culture and therefore have the potential to provide an atmosphere which is supportive of positive behavioural change.

If companies or organisations are prepared to take such initiatives, the question remains as to what they can reasonably expect in return for their efforts. Some might say that there is a moral and ethical argument which transcends all others, and that companies should be prepared to invest in such initiatives purely and simply for the greater good of society and humanity as a whole. The problem is that most financial directors would not be impressed with this argument. They would be looking for a little more reassurance as to what their return on investment is likely to be. In reply to this, I would say that there is a clear relationship between the health and wellbeing of the workforce and their productivity. Not surprisingly, therefore, there is also a strong correlation between worksite health promotion programmes and reduced absenteeism rates, reduced accident rates, improved productivity, reduced health-insurance claims and, in the longer term, reduced illness, premature disability and death. Other benefits, such as improved morale and improved employer/employee relations are more difficult to quantify but no less important. Evidence would suggest that such programmes have a high perceived value among the workforce and can do much to create and to maintain good working attitudes.

We mentioned at the beginning of this chapter that one of the features of change is the fact that new and ever-more-complex skills are required by the workforce so that an organisation can survive and prosper in a rapidly changing environment. Training costs involved in developing such key individuals can be very substantial and therefore the cost of losing them in their prime may be astronomical. Coronary heart disease is perhaps the best example of a disease which kills indiscriminately and which is responsible for a very substantial amount of premature death and disability. For example, the 1985 coronary disease death toll among persons aged 15 to 65 years generated a loss of potential working life estimated at nearly 240,000 years, at a cost to the NHS in England and Wales alone of almost £400 million. The cost to the country as a whole in terms of lost productivity and so on is difficult to calculate accurately, but it is likely to be in excess of £1 billion per annum.

The establishment of a 'Corporate NHS' dedicated primarily to preventive medicine and networked throughout the country represents a major challenge and a major opportunity. It also represents a chance for us to change the emphasis from a narrow disease-orientated view to one which will instead actively promote good health and health maintenance. Not only would it have a major impact on the health of many millions of people in this country, it would also take a considerable amount of pressure off the NHS in the long term. The NHS would then have far more resources available to do what it does best – care for sick people. If corporate organisations in this country were collectively to invest a proportionate share of the amount of money required to modify the risk factors that lead to disease, an astonishingly high payback could be achieved. The active promotion of good health on a proactive basis – rather than a reactive approach to already established disease – is not merely medically desirable. It also makes good economic sense.

CHAPTER 2

THE NORTH AMERICAN AND BRITISH EXPERIENCE

During the past ten to 15 years a revolution in attitudes towards employee health has taken place in many corporate organisations and institutions throughout North America. This revolution finds its origins in rocketing health insurance and medical costs and is fuelled by a growing body of evidence to suggest that company healthcare costs are strongly related to employee lifestyle and behaviour patterns. It seems logical, therefore, that behaviour-related improvements in health should lead to containment of costs, and worksite health promotion or 'wellness' programmes have flourished in response to this. While the actual content may vary as a function of corporate culture, resources and company objectives, these programmes tend to have in common an emphasis on personal responsibility towards health and generally include some form of health screening followed by intervention strategies related to exercise, nutrition, smoking, weight control, stress management and similar risk factors. Enthusiasm for health promotion activities in the workplace derives not only from its intrinsic logical appeal but also from a growing body of professional and scientific evidence supporting the relationship between employee health and economic benefits. As we shall see later, this data is not conclusive, and part of the reason for this is that many such programmes have been in existence for a relatively short period of time.

In general terms, employee health promotion is viewed by most employers from two perspectives. First, many organisations are genuinely concerned for the wellbeing of their employees, and their motivation in offering worksite health promotion may be purely altruistic. Any potential financial benefit for such organisations, is secondary to a concern for their employees and the recognition that such worksite initiatives are well received and valued by the workforce.

The second employer perspective is a purely economic one. In an intensely competitive financial and commercial environment, many employers view health promotion programmes primarily in terms of their potential for saving money. The reasons for this are entirely understandable.

Unlike most companies in the UK, employers in the USA provide healthcare insurance to the vast bulk of employees and their families. In 1988, the premium cost to American employers was $152 billion, providing health insurance to 73 million employees, and 72 million dependants. For many years, neither employer nor employees were particularly concerned about the growing cost of healthcare, but inevitably the financial time bomb exploded. Back in 1960, annual healthcare expenditure in the USA amounted to almost $27 billion, representing just over 5 per cent of gross national product (GNP). By 1980, total healthcare expenditure was approximately $247 billion, running at almost 10 per cent of GNP. In 1985, the total expenditure on healthcare was just over $462 billion and this meant that for the first time in history, Americans began spending more than $1 billion each day on healthcare.

By 1990, the estimated total expenditure in the USA will be in excess of $820 billion and running at about 11 per cent of GNP. According to the Federal Health Care Financing Administration, the national medical care bill in the year 2000, will top a staggering $1 trillion. These trends are indicated in Figure 2.1.

The implications of all this for industry are simply enormous. In 1984, the 'big three' auto companies in the USA, Chrysler, Ford and General Motors, spent a total of $3.2 billion on healthcare. In 1985, Chrysler estimated that its

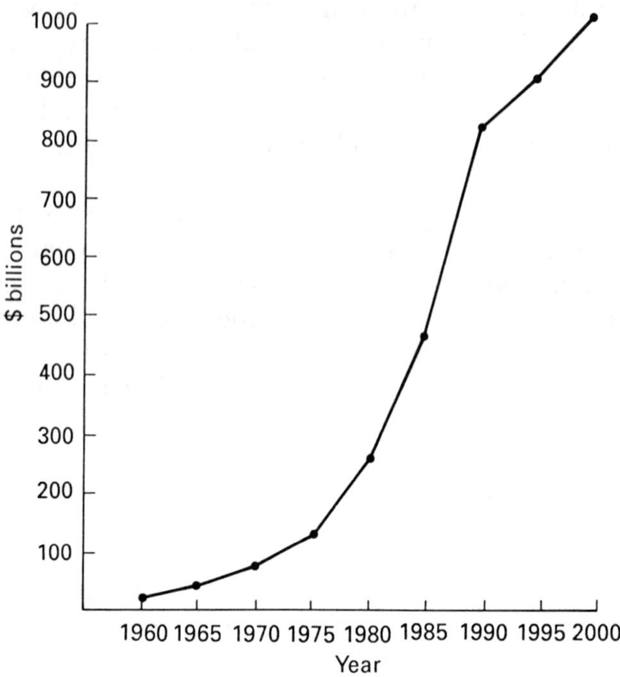

Fig. 2.1 *Trends in USA healthcare expenditure: 1960–2000*

health insurance bill added about $600 to the cost of every car it manufactured. Indeed, the company had to sell about 70,000 vehicles simply to pay for its healthcare bills. Because of this, there has been a growing trend in recent years for American companies to ask their employees for a contribution towards health insurance costs. (See Figure 2.2).

In such a climate it was inevitable that worksite health promotion programmes would be regarded by many employers as one of the most promising strategies to contain their runaway healthcare costs. Indeed for many of them, this was the sole motivating source. This view was made explicit in a recent publication called *Investing in Employee Health: A Guide to Effective Health Promotion in the Workplace.* The authors state that: 'The ultimate goal is to manage healthcare expenditures associated with absenteeism, disability claims

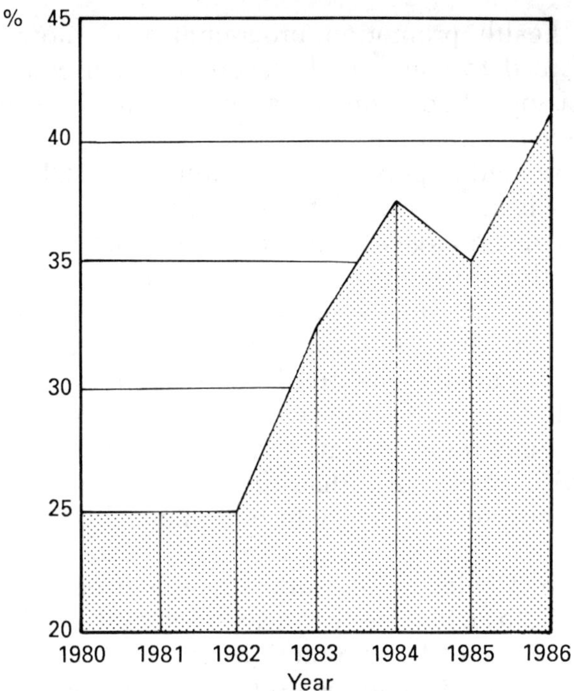

Source: Employee Benefit Research Institute

Fig. 2.2 *Percentage of US workers paying part of health insurance premiums*

and overall insurance costs by developing a healthy and productive employee population.'

No doubt some will be disturbed by such an overtly mercenary attitude and we may well ask whether modern health policy has taken us to the point where we can only consider health in financial terms. Such a question raises profound moral and ethical matters about our culture and civilisation. But if we live in a society which tends increasingly to judge everything in purely financial terms, then it is perhaps not surprising to find health care judged in the same way. The fact is that the decision to implement health promotion programmes resides with senior management, and such decisions increasingly reflect financial investment considerations. Both purchasers and initiators of worksite health promotion programmes are now almost always forced to confront the economic consequences of such decisions. Such considerations are likely to become more, not less, important.

Of course neither perspective – purely financial or purely altruistic – is necessarily right or wrong. An employer who decides to spend large amounts of money on promoting the health of the workforce, at the expense of the financial viability of the organisation as a whole, is obviously not acting in the best interests of the employees or the community. Equally, the employer who pays little or no attention to the health of his workforce may well be undermining their morale and reducing their potential productivity. Not only that, but increasing healthcare costs will add unnecessarily to the financial burden of the organisation.

The correct balance lies somewhere between these two extremes. Both factors are obviously important and decisions on worksite health promotion should involve the consideration of both and the exclusion of neither.

We shall now look at a number of health promotion programmes in the USA, both in terms of their medical effectiveness and also in terms of their potential for cost containment. There are particular problems in trying to evaluate the economic benefits of such programmes and some of these will be discussed in a separate section.

How prevalent are these programmes?

Employer interest in sponsoring health promotion activities began to emerge in the mid-1970s and by the mid-1980s health promotion programmes in the workplace had become commonplace. Until recently, good quality data on the prevalence of such activities were not available. But during the last few years a number of good quaility surveys have been undertaken which provide us with some good indications of how widespread workplace health promotion activities have become. We shall consider three of them in some detail.

The national survey of worksite health promotion activities

In 1985, The US Department of Health and Human Services carried out the national survey of worksite health promotion activities. The primary aim of this survey was to determine the extent and scope of activities in worksites across the United States in private organisations with 50 or more employees. It was also used to assess what employers' perceptions were in terms of the direct and indirect benefits of their efforts to prevent disease and promote employee health in the workplace. This was a large study; the total number of worksites contacted for the survey was 1635 (400 worksites with 50 to 99 employees and 1235 worksites with 100 or more employees). The total number of interviews completed was 1358, representing 83.1 per cent of the total sample.

The survey found that 65.8 per cent of worksites had some form of worksite health promotion activity. The distribution of these activities is shown in Figure 2.3.

It is also quite clear that the prevalence of activities varied significantly as a function of the number of employees at the worksite; the more employees the greater probability that the company would offer a particular programme. This is well illustrated in Table 2.1 below for smoking, health screening, stress management, exercise programmes, nutrition education, blood pressure control and weight control.

The vast majority of organisations made such programmes available to all employees irrespective of status within the company. Indeed, less than 4 per cent of all companies reported that only top management was eligible for such worksite health promotion, and less than 2 per cent reported that only full-time employees were eligible.

More than 80 per cent of all respondents said that they

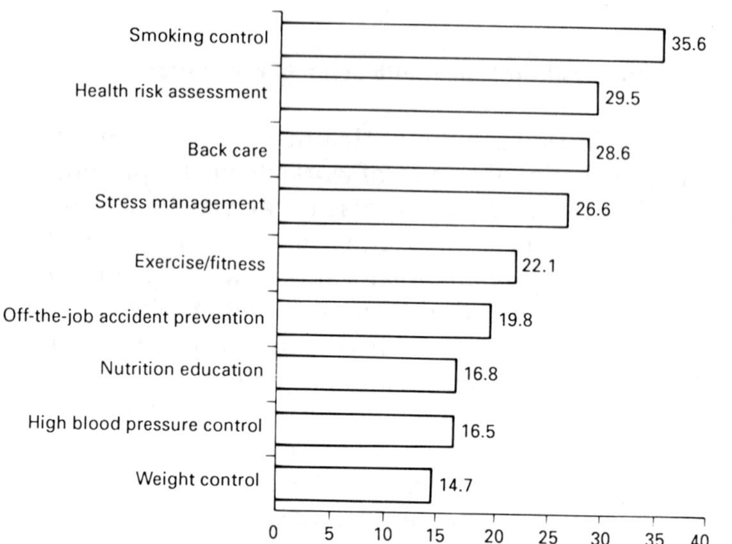

Source: US Department of Health and Human Services

Fig. 2.3 *Prevalence of heath promotion activities by subject (%)*

Table 2.1 *Percentage of worksites offering each type of activity by size of workforce*

	Worksite size (No of employees)			
Type of Activity	**50–99**	**100–249**	**250–749**	**750+**
Smoking control	30.1	37.5	39.5	57.9
Health risk assessment (screening)	18.4	34.0	41.8	66.2
Stress management	14.5	32.7	37.5	60.8
Exercise/fitness	14.5	22.7	32.4	53.7
Nutrition education	8.6	19.8	21.9	48.0
High blood pressure control	8.7	17.9	23.8	49.8
Weight control	8.1	13.5	22.9	48.8

Source: US Department of Health and Human Services

were either extremely or moderately concerned with health-care cost management and an overwhelming majority of respondents with activities indicated that the benefits of their activities outweighed or balanced the costs.

The Health Research Institute's biennial corporate wellness programmes survey

The Health Research Institute has conducted biennial healthcare costs containment surveys since 1979. In 1987 the focus of the survey changed and the purpose was primarily to report on the prevalence and effectiveness of employer actions to control employee healthcare costs through corporate health promotion programmes. Five hundred companies were contacted, selected from Fortunes 500 largest industrial corporations, Fortunes 500 largest service organisations and a number of other employer groups. A total of 144 organisations (28.8 per cent) responded to the survey and this was felt to represent an adequate cross-section of the survey population.

Over 63 per cent of respondents to the 1987 wellness programme survey offered health promotion activities of one form or another. This is nearly 50 per cent more than the proportion of respondents offering wellness programmes in the 1985 survey. The most frequently offered programmes are weight reduction (88.7 per cent) and smoking cessation (86.5 per cent), followed by fitness and stress reduction programmes (78.6 per cent).

Employees eligible to participate in employee wellness programmes covered a wide spectrum. About 90 per cent of the respondents offered the programmes to senior, middle management and blue collar workers. A large proportion (76.4 per cent) also made the programmes available to retired employees.

Companies responding to the survey had a total of almost 3.7 million employees, and the average number of employees per responding company was around 26,000.

As we have seen from the previous survey, the prevalence of such programmes tends to be much higher in larger companies and this survey also supports that finding.

Corporate Health Care Management Company/health promotion programmes survey

In 1986 the Corporate Health Care Management Company conducted a market study regarding the prevalence, nature and potential of employer-sponsored health promotion programmes in the USA. Two hundred telephone interviews were conducted with decision makers responsible for employee benefits of US companies with 500 or more employees. Health promotion programmes – defined as 'programmes designed to educate, motivate, and help employees to adopt and sustain healthy lifestyles – were used by 51 per cent of the respondents. Of these, 35 per cent reported that these activities had been in place for at least one year. Companies reported activities in the following areas:

- 80 per cent employee health education programmes
- 40 per cent health screening
- 37 per cent smoking cessation
- 25 per cent nutrition and weight loss
- 23 per cent exercise
- 14 per cent stress management.

Some of the overall reactions to these programmes are worth mentioning: 97 per cent of the companies believed that healthy, fitter employees have fewer health insurance claims; 94 per cent indicated that the general good health of their employees is an important corporate responsibility. In 85 per cent of cases senior management believed that health promotion activities of one form or another would help to reduce medical expenses in the long run.

These surveys show that worksite health promotion is a common feature of American organisations, with between half to two thirds offering activities of one kind or another. However, a closer examination of the evidence reveals that *comprehensive* programmes which involves management of multiple disease risk factors and a combination of intervention strategies, are *not* common. Furthermore, those comprehensive programmes which do exist are found overwhelmingly in the larger companies.

Some programme examples

Johnson & Johnson

Johnson & Johnson, based in New Brunswick, New Jersey, has one of the most often quoted and best documented health promotion programmes in the United States. The Live for Life (LFL) programme started in 1979 and had two primary aims: to provide the means for Johnson & Johnson employees to become as healthy as possible and to determine the extent to which the programme could be shown to be cost effective.

The programme itself consists of an initial health risk appraisal which is carrried out using questionnaires covering lifestyle issues such as alcohol and cigarette consumption, nutritional and exercise habits, and so on. This is followed by a ten-minute physical examination which includes measurement of blood pressure. Then there is a one-to-one health counselling session where specific issues relating to risk factor modification are discussed in more detail and prescriptive measures are recommended depending upon the findings of the initial assessment. Employees can then go on to participate in a variety of programmes designed to bring about improvements in general lifestyle and behaviour patterns.

There is good evidence that this programme does produce significant changes in overall behaviour patterns which have undoubtedly improved the overall health of the workforce. LFL employees reported 95 fewer sick days than in the year before the programme started. Those employees without the

LFL programme reported a 13 per cent increase in sick days. Compared with non-LFL sites, LFL sites had significant reductions in smoking, weight, and stress levels and improvement in fitness and exercise levels. For example, 23 per cent of the workers who smoked at LFL companies in 1980 were no longer smoking in 1982; this compares with an 8 per cent reduction at companies without the programme.

From the cost containment point of view, an analysis of the data has been carried out by researchers at the University of Pennsylvania and the Research Triangle Institute. The study involved three groups: two groups of approximately 5200 and 3300 people respectively who were participants in the LFL programme and a third group of about 3000 people who were non-participants in the LFL programme. The mean annual inpatient cost increases for both the LFL groups and for the control group were compared over a five-year period. Mean annual inpatient costs for the two LFL groups increased by $43 and $42 respectively, while the cost for the control group increased by $76. The number of mean annual hospital days per 1000 employers were 109 and 67.5 respectively for the LFL groups and 171 for the third group. The researchers concluded that hospital costs for the LFL groups had doubled over the five-year period while costs in the non-LFL group had increased four-fold. (See Figure 2.4.) This translated into savings for those companies offering the LFL programme of almost $250,000 per year (expressed in 1979 dollars).

Control Data Corp

Control Data Corp of Minneapolis initiated its Stay Well programme in 1979. The programme was initially developed for Control Data's employees but today it is also available to other organisations through a nationwide network of over 50 authorised distributors. Stay Well is extremely comprehensive, and the company has also made a major

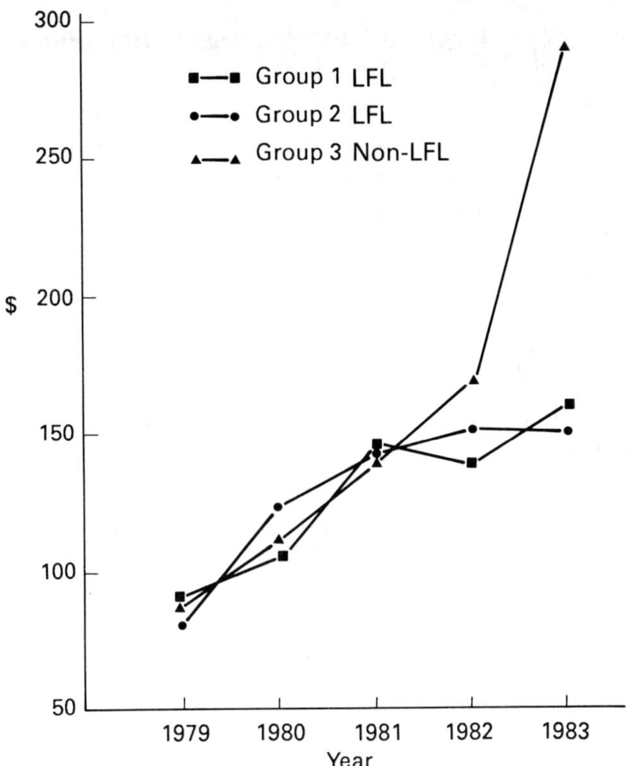

Source: Bly et al, Journal of the American Medical Association

Fig. 2.4 *Per capita inpatient costs 1979 – constant dollars*

commitment to evaluating all aspects of the programme including its medical impact and its cost containment potential.

Perhaps one of the most impressive features of the programme as a whole is the meticulous attention to detail and the careful planning and assessment which take place prior to any workplace initiative being established. The programme has been phased into more than 100 facilities and about 70 per cent of eligible full-time employees now participate.

The process starts with a general employee health survey followed by careful site planning which prepares the worksite and key management for successful implementation. This is followed by specific orientation sessions designed to explain the programme to prospective participants. At the commencement of the programme proper the individual undergoes a health screening examination designed to identify any major disease risk factors and this information together with the results of any blood investigations is fed into a computer which then produces a specific health risk profile. By participating in programmes related to smoking cessation, weight control, fitness, blood pressure control, relaxation techniques and nutrition, the employee is encouraged to improve his risk profile as much as possible.

In addition to this, the programme itself is supported by specific courses that focus intensively on a particular health issue, and there is also a quarterly news magazine for employees and their families. Specific courses to address risk factors are available in:

- Fitness
- Smoking cessation
- Stress management
- Nutrition
- Weight reduction
- Back care

Each course and programme component has been designed so that it can be used on its own. This ensures the flexibility necessary to address specific individual needs. Evidence suggests, however, that combining components into a highly visible and comprehensive health promotion programme creates more overall impact than would otherwise result.

The Stay Well programme has been well received by Control Data employees and major lifestyle changes have taken place within the employee population. There is a particularly encouraging trend among those employees who were originally classified as being in high risk groups. Participation by these employees in relevant programmes has reached 54 per cent for overweight, 46 per cent for hypertension, 45 per cent for cholesterol, 33 per cent for fitness and 23 per cent for smoking.

Risk in two key areas, smoking and weight, has been reduced to a significantly greater degree at sites where the entire Stay Well programme is available compared with other worksites where the Stay Well programme has not yet been fully implemented. (See Figures 2.5 and 2.6.)

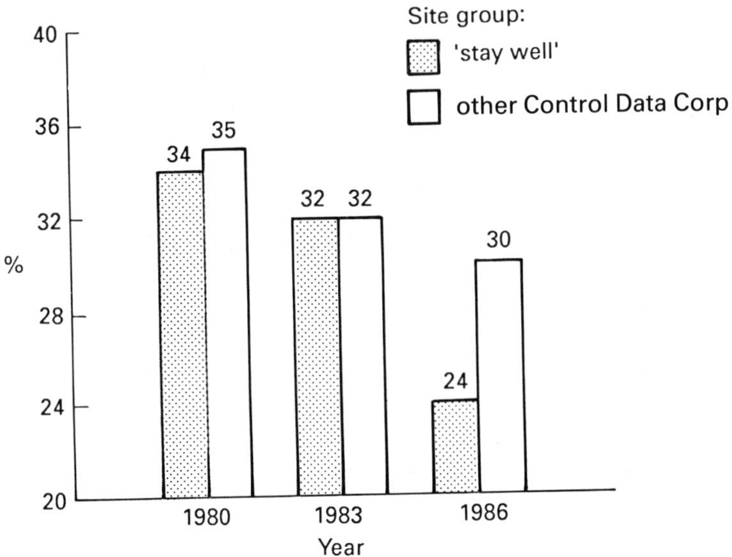

Source: National Wellness Institute

Fig. 2.5 *Smoking trends (per cent smoking cigarettes)*

To provide some realistic assessment of the cost containment potential of these workplace strategies, it was necessary to construct a model of projected healthcare costs for a high versus a low risk male. A high risk male was defined as one who smoked two packets of cigarettes a day, was sedentary, was at least 30 per cent overweight, had a high cholesterol level and did not use seat belts. The low risk male was a non-smoker who exercised moderately, was not overweight, had a low cholesterol level and wore seat belts most of the time. The difference in projected healthcare costs for these two

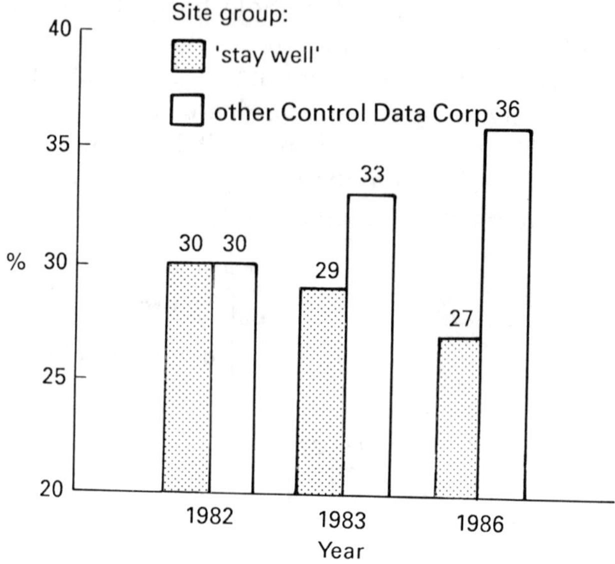

Source: National Wellness Institute

Fig. 2.6 *Weight trends (per cent in overweight range)*

individuals is indicated in Figure 2.7.

By comparing the estimated excess cost of a high risk individual from this model with information about reductions in employee risk over the period from 1980 to 1986, it was possible to estimate the savings in healthcare claims and absenteeism due to risk reduction. The estimated annual saving in healthcare claims was approximately $1.8 million. The authors of the study were, however, cautious about interpreting this estimate. They did not consider that the evidence of this study alone was sufficient proof that the cost of the health promotion programme was less than the benefit derived from it. Neverthless, these data are clearly encouraging.

Tenneco Inc

Tenneco Inc started a corporate fitness and health programme in 1982 based at its Employee Centre, and $11 million facility

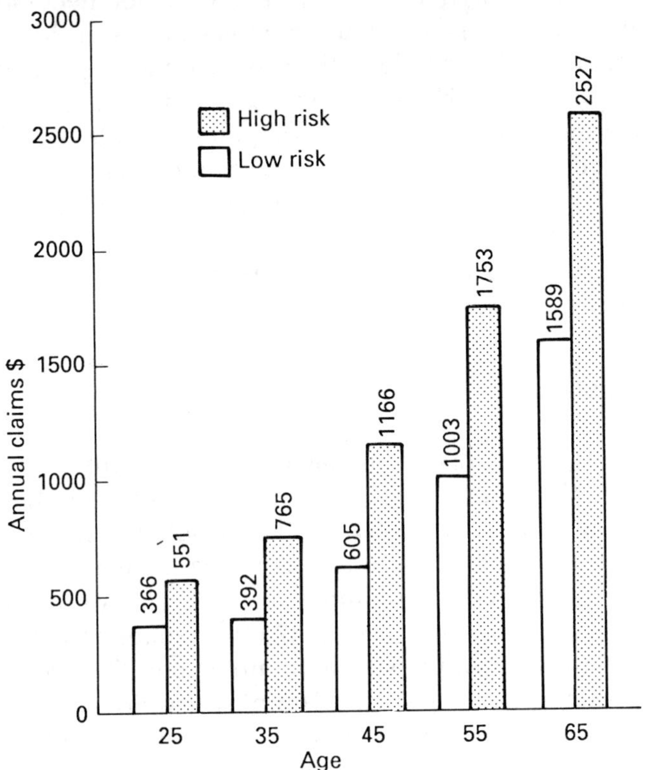

Source: National Wellness Institute

Fig. 2.7 *Annual healthcare claims, low vs high risk males (1984 dollars)*

in the corporation's Houston headquarters. The permanent staff of 11 individuals are mostly exercise physiologists, and the equipment includes weight-training machines, running track, four racket ball courts, whirlpool and sauna. The emphasis is on exercise, but the programme also has specific programmes for smoking cessation and blood pressure control.

Tenneco has found a strong correlation between exercise and work performance, but is well aware that a person who has the discipline to exercise regularly also tends to have the attributes of a high performer. There is not necessarily a causal relationship between the programme and the level of performance. Nevertheless, there is evidence to suggest that the programme is beneficial in cost containment. In 1984, average medical costs to the company of male and female non-exercisers were $442 and $896 higher respectively than those of male and female exercisers. Another study of 3200 Tenneco workers found a positive correlation between exercise and productivity. Better quality data are likely to be available in the future.

Kimberly Clark

Kimberly Clark's health management programme was initiated with three aims in mind: preventing illness in the workforce, improving productivity and reducing absenteeism. It is based at its Neenah, Wisconsin headquarters in a $2.5 million facility which includes an 8000 square foot multiphasic screening unit and a 37,000 square foot exercise facility with an olympic-size swimming pool. The programme itself began in 1977 for salaried employees and was extended to other workers in 1979 and to spouses and retired workers in 1982. There are four major elements to the programme:

- Medical screening
- A fitness programme with an inhouse cardiac rehabilitation unit
- A health-education programme
- An employee assistance programme covering alcohol and drug abuse as well as mental problems.

The company expected that it would take at least a decade before any reduction in claims or absenteeism could be demonstrated, and there is not yet any documented evidence of the programme's effects on healthcare costs. However, the screening programme has detected a number of employees at high risk for high blood pressure (hypertension), heart attack and cancer. In addition, the company has found the programme of great benefit in recruiting and retaining high-calibre employees.

Blue Cross and Blue Shield

Blue Cross and Blue Shield of Indiana started their Stay Alive and Well programme in 1978. It consists of three main phases:

- There is an initial physical examination (including measurement of height, weight, blood pressure and plasma cholesterol) and medical history.
- On the basis of these initial results, the individual is given specific information as to which particular risk factors are prevalent and how these can best be modified.
- The third phase involves specific health risk intervention programmes addressing such issues as correct diet, smoking cessation and weight control.

Between 1978 and 1986 more than 2100 of the total of 2500 employees participated in the programme. More than 600 employees attended nutrition classes, more than 300 employees went through some form of weight reduction (weight loss of more than one ton was logged!), and more than 300 employees attended smoking cessation groups. Half of these people either stopped smoking entirely or made big reductions in their cigarette consumption.

In 1980 the Kellogg Foundation helped to fund a $240,000 research project designed to test the effectiveness of the programme in a number of ways. Two questions in particular were addressed:

1. Did the programme reduce the prevalence of disease risk factors within the employee population?
2. Did the money saved through reduced absenteeism and health care claims justify the programme from a financial point of view?

The answer to each question appears to have been an unequivocal *Yes*. First the study showed that employees participating in the Stay Alive and Well programme significantly reduced their weight, their cholesterol levels and their blood pressure. These changes were monitored over a five-year period and appeared to be permanent. In addition, absenteeism rates among participants dropped anywhere between 50 and 400 per cent depending upon the individual.

Blue Cross and Blue Shield spent about $867,000 over a five-year period, but saved a total of $1.6 million. These data provide further evidence that health promotion programmes can reduce health risk factors, absenteeism and healthcare costs.

United Methodist Publishing House (UMPH)

The United Methodist Publishing House, an organisation with some 1300 employees, has implemented a comprehensive programme for health enhancement and accident prevention. The programme began in November 1984, and is called Try UMPH for Health. It was developed to help to control the company's medical claims costs, which increased at an average of 16.5 per cent a year between 1980 and 1984. An unused part of the printing plant was converted to a fitness facility and a comprehensive system of health checks, stress, smoking, nutrition and weight control programmes were made available. More than 600 employees participated in some component of the health programme in the first 18 months.

In order to measure the cost effectiveness of the programme UMPH made a number of comparisons. First they compared the total medical claims costs for 1984 when the programme was commenced, with those for 1985. The health benefit plans were similar for both these years and so the comparison was felt to be a reasonable one. According to the US Health Care Financing Administration, national healthcare costs grew approximately 9.4 per cent in 1985. However, the total medical claims costs for all UMPH employees *decreased* in 1985. If UMPH costs had risen in line with the national average, the 1985 figure would have been just over $2.2 million. In fact it was just over $1.82 milllion, representing a saving of more than $375,000. Of course, this may just have been a particularly good year but UMPH intend to follow this trend carefully to establish whether or not the programme itself contributed significantly to these cost savings.

UMPH also compared 1985 medical claims costs for participants and non-participants in the programme. The results, shown in Table 2.2, suggest that individuals who participate in health promotion programmes are healthier and may therefore be expected to cost the company less in healthcare costs than non-participants.

Table 2.2 *Average UMPH medical claims per employee*

	Programme participants	Non-participants	UMPH savings
Female	$320	$875	$555
Male	$267	$469	$202

Nevertheless, it is important to recognise that these figures are based only on one year's experience and may not be very meaningful. Careful monitoring over a number of years will yield important medical and financial information on which to base more rigorous analysis of the programme's long-term effects.

Aldoph Coors Co

Aldoph Coors, the Colorado-based brewer and manufacturer, began promoting recreational programmes for its workers some 35 years ago. In the early 1970s, an additional employee assistance programme (EAP) was set up to help workers deal with alcohol and substance abuse and mental problems. In June 1981, Coors opened a 23,000 square foot fitness facility near its headquarters in order to encourage its workers to become responsible for their lifestyles and general behaviour patterns. Off-site fitness facilities are now available to workers at eight Coors subsidaries.

The Coors health promotion programmes are made available to employees, their spouses, dependants aged 12 years and above, and to retired workers. About 20 to 25 per cent of those eligible do participate actively. Before they start a programme employees undertake a health hazard appraisal which pays specific attention to such lifestyle factors as smoking, excess weight and inactivity. Each individual receives a specific health prescription, based upon this data, which assists him in reducing his risk profile to a minimum. Some 5 per cent of the participants are identified as being at particular risk for cardiovascular disease and they are offered treadmill exercise stress tests. After the initial health assessment, participants can choose from six risk factor modules. These are physical fitness, smoking cessation, weight loss, nutrition, alcohol and stress management. There is good evidence that the initial health assessments and the subsequent intervention programmes have produced measurable benefits in risk reduction.

The Coors organisation has also kept track of the economic payback that these programmes could provide. On the assumptions (based on studies conducted by other companies) that people who exercise regularly receive 60 per cent less health insurance benefits and are absent from work about 40 per cent less often than non-exercisers and that the company will save about $200 annually for each worker who stops smoking, Coors estimated that in 1984 for each dollar spent on the health promotion programme, it received $2.38 return on investment.

A T & T Communications

The A T & T Communications programme, Total Life Concept (TLC), was piloted in 1983 and initially involved employees at nine sites participating in an evaluation study. The evidence was very promising and suggested that significant reductions in risk of premature disability from heart disease and other conditions could be brought about by the programme.

A T & T used two control groups in assessing its health promotion programme. One group was given no information and no health risk appraisal, while the second was given a health risk appraisal but no follow-up information in terms of specific education programmes. On the other hand, the study group itself received both an initial health risk appraisal and subsequent risk factor modification by various health education modules in areas such as weight control, fitness and smoking cessation.

Those in the study group showed an overall improvement compared with the control groups. They exercised more, they smoked less and they had an improved perception of their own health. A T & T estimated a long-term saving based on the improvement in the study group's health as compared with the control groups. By extrapolating from this study to the entire employee population of 110,000 A T & T predicted that no fewer than 374 heart attacks could potentially be prevented over a ten-year period. If this occurred, it would result in savings to the company of about $93 million in direct and indirect costs. In addition, the company estimated potential savings of $17 million over the ten-year period if cancer prevention results were similarly generalised.

Health promotion in smaller companies

Most of the impetus for the development of worksite health promotion programmes has come from larger corporations.

However, workplace health initiatives are by no means confined to big corporations; a growing number of small and medium-sized companies also recognise the value of prevention and health education in the workplace.

Scherer Brothers Lumber, a sawmill firm based in Minnesota, has developed a health promotion programme which is beginning to yield some positive results after seven or eight years. Initially they set up a health promotion committee made up of members of management and employees and held a series of seminars on weight reduction, smoking cessation and blood pressure. This relatively simple programme eventually expanded to include programmes designed to reduce alcohol abuse, absenteeism and accident rates. The company reports significant reductions in absenteeism and accidents as well as growing savings on health insurance and workers compensation premiums. The initial start-up costs for the programme were relatively high and the financial returns were at first minimal. However, after seven to eight years, the company estimates that the programme is becoming profitable. In 1987, the annual costs for the programme were approximately $107,000, but savings in health insurance, absenteeism and workers compensation insurance amounted to about $204,000. The net annual savings were therefore estimated to be approximately $97,000.

The Safeway Bakery Division in Clackamas, Oregon, with 120 employees, has invested in a health maintenance facility and provides health screening and specific programmes related to nutrition, stress management, safety at work, and so on. No detailed cost studies are available but the programme is said to have resulted in significant savings from reduced absenteeism, staff turnover and industrial accidents.

The Times Publishing Company in St Petersburgh, Florida, which owns the St Petersburgh Times and several affiliated companies, started a health promotion programme in 1982. The programme is available to all employees, retired workers and family members and is made up of a health screening programme supported by a series of preventive health initiatives and intervention strategies. These include the standard components of smoking cessation, weight reduction, stress management and exercise programmes. The prime motivator for the initiation of the programme was the spiralling health insurance costs which were having a very significant impact upon the financial performance of the organisation. In the five years prior to the commencement of the programme, health insurance claims had risen by 90 per cent. The company estimates that in the first 18 months of the programme they experienced a saving of some $200,000 in health insurance costs.

Employee assistance programmes (EAPs)

Employee assistance programmes offer counselling and referral services for workers and dependants suffering from psychiatric disorders, alcoholism and other forms of drug abuse and also financial, legal and marital problems. In some senses, therefore, they can be regarded as a parallel and complementary activity to worksite health promotion programmes.

Most of these schemes are operated either in-house by the employer, or else under contract to an outside organisation which provides these services. As with most health promotion programmes, EAPs and the services they provide are paid for almost entirely by the employer. Participation is voluntary and strict confidentiality is an absolute prerequisite for success. Experience suggests that most problems for which EAPs are consulted, can be dealt with in one or two sessions of counselling. Those employees with more intractible problems which cannot be resolved so easily, are usually referred to appropriate healthcare providers and community services.

EAPs originate from a programme established by Macy's

department store in 1917 to help employees with personal problems. At about the same time organisations such as the Metropolitan Life Insurance Company and Western Electric also established similar schemes. By 1958, more than 50 companies had occupational alcoholism programmes. Today there are several thousand EAPs operating throughout the United States.

While start-up costs for EAPs can be rather high, advocates argue that they ultimately save employers considerable amounts of money be resolving many problems at an early stage and thus avoiding the need for long-term treatment with all the associated costs. For example, A T & T Communications reports that an EAP saves $2.59 for every $1 invested. United Airlines is said to have a return of $16.35 for each $1 invested in its own EAP. However, many people remain sceptical as to the ability of EAPs to help control healthcare costs. The major problem appears to be a lack of methodologically and empirically sound research data, on which to base meaningful conclusions. This, as we shall see later, has also been a problem for cost containment arguments in relation to worksite health promotion activities.

The British experience

Thus far I have concentrated exclusively on the American experience but it would be wrong to give the impression that nothing is happening in UK companies. In fact there are quite a number of exciting and promising initiatives presently underway. Moreover, many companies and organisations believe that these health promotion and screening programmes *are* returning significant economic benefits to the organisation, especially when considered in the longer term. The problem is that this information has not been published and can be described as only anecdotal evidence. Moreover, it is likely to remain that way in most cases because there is very little incentive for an organisation to spend time and money on formal studies. The acid test is, after all, whether the programme works or not and if it does not, then most comapanies will simply stop doing it. Nevertheless, the fact that there is growing interest in worksite health promotion among corporate organisations in the UK suggests that many of them believe that significant benefits are to be obtained, especially in reduced absenteeism, improved employee morale, improved productivity and reduced sickness rates. Although such benefits have not yet been quantified scientifically, existing evidence provides a powerful incentive for companies to examine the potential advantages of worksite health promotion activities.

Before outlining some of the programmes currently in operation in the UK, I should point out that many companies have a genuinely altruistic interest in the health of their employees and for them producing figures to support financial returns is not important.

Marks & Spencer

For many years Marks & Spencer has been regarded as setting the 'gold standard' levels in health provision for employees. The company has an extensive occupational health programme which began in earnest in the 1950s and was aimed at promoting the health and wellbeing of the employees in the workplace. The programme is very extensive and is made available to all employees in the organisation. In recent years the programme has expanded considerably and now includes a number of specific screening programmes and health promotion initiatives.

The cervical cancer screening programme began in 1968 and is available to all female employees on a three-yearly basis. Breast cancer screening, including mammography, commenced in 1973 and is available to all women of 40 years of age and over every 18 months. This is carried out using the mobile services of a private medical organisation. The company also offers health checks to managers aged 30 and over on a three-, two-, or one-yearly basis, depending on age.

Marks & Spencer has now embarked upon an ambitious screening programme designed to identify risk factors for coronary disease in all employees aged 30 and over; it is supported by various intervention strategies aimed at encouraging healthier eating, smoking, exercise and drinking habits. The screening examination is carried out on site in the various M & S stores throughout the country and is administered by a highly trained and experienced nurse. Using a portable computer, the nurse goes through a comprehensive health questionnaire and carries out a number of specific measurements including height, weight and blood pressure. A blood sample is then taken which is sent to a central laboratory for detailed analysis and the screening examination concludes with a brief counselling session supported by appropriate written information. This is followed up at a later stage by store visits from members of the Health Education Authority who reinforce the health education message delivered at the time of the initial screening.

During the next three to five years, 30,000 to 40,000 employees (90 per cent of whom are female), in more than 270 locations throughout the UK, will be invited to participate. I have personally been closely involved in the design and implementation of this project, which I believe to be of considerable importance. Not only is it likely to have a significant impact upon the behaviour and future health of M & S employees, it will also provide much needed research data about the origins of coronary disease in women. This is particularly relevant because there is growing evidence that heart disease in women is on the increase, especially in the younger age groups. At the present time, the pilot study involving more than 2000 employees has been completed, and the organisation will shortly be moving into the main phase of the project.

Marks & Spencer runs a variety of other smaller-scale

programmes, including a campaign to encourage regular self-screening for testicular cancer in males, colorectal cancer screening and some more recent work examining the possibility of screening for skin cancer. The medical department is also evaluating the use of abdominal ultrasound in the diagnosis of ovarian cancer and some forms of arterial disease.

By any standards, this enormous range of services represents a unique commitment on the part of M & S to the health of its employees. The philosophy underpinning their approach is summarised by their Chief Medical Officer, Dr Derek Taylor:

> We apply the mixed economy in our approach to health promotion, which I would define as helping people improve their health through *education* – raising the level of awareness on how to achieve better health through lifestyle modification; *screening* to detect factors for disease or detect disease itself at an early stage in order to facilitate prevention or more effective treatment; thirdly, *action programmes* – to put the theory of risk factor prevention and management into practice.

The Post Office

The Post Office employs in excess of 200,000 people, so clearly any health promotion activities for the workforce have considerable financial and logistic implications. Apart from providing the more traditional occupational health services, the medical department has also undertaken a number of major initiatives in health promotion and disease prevention. The view is that these programmes will bring about a significant improvement in the health of the workforce, improve employee relations, reduce absenteeism and be an effective means of breaking down the traditional barriers between management and workforce. In the longer term, it is hoped that significant economic benefit will also be demonstrated.

A breast and cervical cancer screening programme was launched in 1988, which was made available to employees at various worksite locations throughout the UK. The screening examinations are carried out in a specially commissioned and equipped mobile unit and, during the next two years, 30,000 to 35,000 women are expected to use this service. I have worked closely with the Chief Medical Officer, Dr Richard Welch, and a number of specialist advisers in the design and organisation of this programme. Initial results appear very encouraging.

The Post Office runs two mobile health promotion units, the first commissioned in 1986 and the second in February 1989. These initiatives have also been extremely successful.

The organisation has also been active in the area of stress management and now employs two specialist counsellors who offer in-house training programmes in stress counselling techniques for occupational health nurses.

A small number of blood testing centres primarily concerned with cholesterol are currently undergoing evaluation, and this Post Office service is likely to be extended throughout the company.

A testicular cancer prevention programme has also been shown to be extremely effective; of the first 300 employees examined, 10 required further investigations.

The Occupational Health Department designed an alcohol abuse programme which was introduced in 1983. This, too, is proving successful and is supported by an excellent in-house video on the subject.

More recently the Post Office has embarked upon a major effort to improve the fitness of the workforce by encouraging more employees to take regular exercise.

IBM (International Business Machines)

IBM has made a sizeable investment in improving the health of its workforce and since 1986 has provided a comprehensive health screening programme for its 18,000 UK staff.

The health screening examination includes blood, lung and ECG tests together with cervical smears and breast examination for female employees. All these are offered free of charge.

The programme revealed that no less than one third of the workforce were found to have a blood cholesterol level representing an increased risk of heart disease, and the overwhelming majority of both males and females were found to be both overweight and unfit. About 15 per cent of males and 23 per cent of females were smokers.

The cost of the programme was estimated at about £70 per head, but this appears to have been a worthwhile investment. Absenteeism rates in certain parts of the organisation have already been reduced from 5.7 to 4.8 per cent.

Polaroid

The Vale of Leven Health Promotion Project was started in 1986 for the workforce at Polaroid UK, based in Dumbarton in Scotland. The essence of the programme is that it should be a community health promotion project and it is funded primarily by Urban Aid money, of which three-quarters comes from the government and one quarter from the local authority. It is also funded by the Scottish Health Education Group and, of course, by Polaroid itself.

Polaroid has approximately 1400 employees, of whom about 87 per cent have participated in the programme. In general terms, the programme involves initial health assessments, including cholesterol measurements, followed by counselling and various intervention strategies. Initial evidence would suggest that significant changes have occurred in cardiac risk factors, especially in blood cholesterol levels. Other risk factor changes are currently being assessed.

The company estimates that to screen approximately 1500 people on the worksite costs about £14 per head. No detailed cost benefit analysis is being carried out because cost was not the main reason why the programme was established in the first place.

The most exciting and interesting thing about the project is that because it was primarily a *community* health project those recipients of the programme on the site at Polaroid are now going into the community and 'spreading the gospel'. A community exercise group has already been set up and other programmes are likely to follow. Local people who wish to participate are charged a nominal fee of 50 pence per session.

The project employs four people; a physical training instructor, a dietician, a secretary and a social scientist. These are supported by the Chief Medical Officer and two nurses.

This is an excellent example of how workplace health enhancement programmes can be extended into the local community, thus achieving a much wider degree of participation. This is obviously good for the local population but also adds considerably to the standing of Polaroid in the local community.

Ethicon Ltd

Ethicon Ltd, a subsidiary of the Johnson & Johnson organisation, is a manufacturing company employing approximately 1500 people at four factory sites in and around Edinburgh. The company currently offers a worksite health promotion programme based upon the Live for Life programme of Johnson & Johnson in the USA. An initial health assessment and screening programme is supported by a number of specific intervention programmes addressing common risk factors such as cigarette smoking, poor dietary habits, obesity and inactivity. The programme was introduced in 1984.

There are no data currently available on either the medical efficacy or the cost effectiveness of the programme. However, the organisation intends that this information should be collected and analysed scientifically, and so important information should be forthcoming over the next couple of years or so.

Economic implications of worksite health promotion programmes

Despite genuine altruism on the part of many companies, the fact remains that in the current economic climate, net cost savings are the strongest argument for continuing and expanding worksite health programmes. Decision makers within organisations today are being confronted by a large and growing body of literature proclaiming the cost effectiveness of worksite health initiatives. Much of this literature has in the past been strong on enthusiasm and rather low on proven fact; the information has been unscientific and could be dismissed as merely anecdotal. But this situation is improving. More recent studies, including many of those reviewed in the preceding pages, are of much tighter design and have yielded better quality evidence which is by no means conclusive but is extremely encouraging. An objective assessment of the economic impact of these programmes is predicated upon the existence of a body of knowledge which is scientifically sound. I am not saying that potential benefits do not exist, but merely that with a few exceptions (referred to later), the current data do not permit any conclusive judgment on the issue.

It does no one any good to make exaggerated claims based upon inadequate evidence. I believe that this tendency has mitigated strongly against the setting up of good, quality, long-term studies from which more definitive evidence could be obtained.

While accepting that good scientific studies are essential, we should not become so obsessed with published data that we *totally* disregard all other evidence. After all, many companies do not feel compelled to go to the trouble and expense of having their evidence scientifically validated; it is sufficient for them to know that the programme works from an economic point of view as defined by their own internal criteria. So the lack of high quality scientific evidence does not in itself invalidate the view that worksite health promotion activities make good financial sense. Corporate organisations are not scientific institutions, but they are usually very shrewd at making financial judgements. The fact that such programmes have now become so widespread, particularly in the United States, suggests that many directors and senior executives are quite satisfied with the evidence even if some scientists are not.

Limitations of methods

Few organisations have the necessary resources to carry out methodologically sound and highly complex research on a long-term basis, nor do they have the necessary background to analyse and assess the data. These inherent limitations have produced serious flaws which are common to most of the studies carried out to date. The more important deficiencies are as follows:

- Lack of sufficiently specific definition or measurement of risk factors, or interventions. For example, what exactly is meant by 'sedentary lifestyle' or 'stress manangement'.
- The tendency to use levels of participation as an index of programme success. This is clearly inappropriate. It is changes in *behaviour* which matter, not merely levels of participation.
- Too much reliance on self-reporting of programme participants, as an index of behavioural change. Self-reporting is a notoriously unreliable method of quantifying the extent of behavioural change within any population. Participants tend to report what they think the study organisers want to hear.

- The absence of good baseline measurements against which future comparisons can be made.
- The absence of control groups.
- An unstable population. Some organisations have a high staff turnover and this mitigates strongly against meaningful long-term comparisons.
- The failure to take into account external factors which may have a significant effect on behavioural changes within the organisation.
- The failure to take into account self-selection. Some individuals who are already health conscious may self-select for such programmes and behavioural changes in this group may not be typical of behavioural changes in other less compliant groups.
- Inadequate sample size and lack of long-term follow-up.
- A bias has often been introduced because those individuals who initiate and run the programmes, are often those who carry out the evaluation of the benefits obtained. There is therefore (not surprisingly!) a tendency to report positively, rather than negatively.

More recent evidence – some encouraging trends

Despite all these difficulties some recent studies have been exceptional. Johnson & Johnson, Control Data and Blue Cross/Blue Shield, which we have already discussed, are examples of organisations which have attempted a much more rigorous empirical and methodological evaluation of their programmes. For this reason their results stand up to close scrutiny, and all three demonstrate significant medical and economic benefits. The Johnson & Johnson study was particularly impressive, not only because of its overall design, but also because the subsequent analysis of the data was carried out by an independent group – the Research Triangle Institute. Even these studies cannot be regarded as being conclusive, a point which the authors themselves were ready to admit. However, we should not underestimate their importance not only because of their positive findings, but (and perhaps more importantly) because they set a standard and an example which one hopes will encourage still higher standards of research and evaluation.

In addition to the economic evaluation of comprehensive programmes such as the ones described above, other studies on the economic benefits of individual health promotion strategies are beginning to yield positive results. Current evidence on smoking cessation and hypertension (blood pressure) control is especially encouraging and once again marks an important and significant trend.

Properly designed and methodologically sound long-term research programmes are difficult and costly to mount. There is a 'catch 22' situation here. Those companies that develop programmes primarily to save money will surely be reluctant to pay out large amounts to have their data analysed. On the other hand, those companies and organisations who initiate worksite health promotion programmes for other reasons are also not going to spend money on any detailed financial evaluation, because that is not their primary interest.

The way forward

Whether the primary motive in setting up a worksite health promotion programme is financial, or is rather a genuine desire to improve the health and wellbeing of the workforce, some means of evaluating the impact of the programme is essential. Reliable baseline data are fundamental to this process and provide a means of evaluating the success and overall impact of the programme in relation to predefined goals.

Cost effectiveness, cost benefit and return on investment are three distinct but related measures of economic impact. A cost effectiveness analysis estimates the benefit of a programme in health terms such as the extent to which blood pressure reduction, cholesterol reduction or smoking cessation is achieved in relationship to the total investment. It is really a measure of the medical effectiveness of the programme in relationship to the total medical cost.

Cost benefit analysis takes a rather broader view and attempts to translate these changes in the medical status of the workforce into financial payoff. In other words, because the workforce is healthier, there should be less premature death and disability; cost benefit analysis attempts to quantify this in hard financial terms. This is often extremely complex and difficult to perform.

The third approach, return on investment, is the more familiar method of evaluating such programmes. It relates the total cost of the programme (direct and indirect), to savings in medical care claims, disability or workers compensation payments, reduced recruitment and retraining expenses or increased productivity.

The harvesting of high quality baseline data is also a prerequisite for the accurate evaluation of the programme from a clinical point of view. Thus measurements of blood pressure, cholesterol levels, smoking habits, absenteeism rates and any other relevant factors need to be carefully organised. The importance of these initial measurements cannot be overestimated since there is nothing worse than spending four or five years running a programme only to find that the data are next to worthless because the baseline measurements were inadequate.

It is also important for us to be able to make meaningful comparisons between one worksite programme and another, whether from similar or from different industries. If standardised methods of benefit and cost analysis were to be adopted, then realistic comparisons could be made allowing the organisations with less successful programmes to learn from their more successful counterparts.

The USA versus the UK

In discussing the potential economic benefits of worksite health programmes, I have focused exclusively on the North American experience rather than that in the UK. The reason for this is quite simple: virtually all the studies on the economic impact of such programmes have been carried out in the USA. No similar body of data exists in the UK and it is likely to be many years before we are in a position to draw conclusions from our own recorded experience. Employers in America have a greater incentive to invest and evaluate programmes because of the nature of their national healthcare system. In the USA, the employer, more often than not, foots the bill for the health insurance of his employees; in Great Britain, this is the exception rather than the rule.

This fundamental difference makes it almost impossible to draw any conclusions regarding the potential economic advantages of such programmes in the UK. Unless the background circumstances are identical, or at least very similar, any extrapolation from the North American experience to that in the UK, is inappropriate.

However, current trends would strongly suggest that this situation is likely to change over the next few years. Employers are becoming much more disposed to provide health insurance for their employees, and increasingly it is for the *whole* of the workforce rather than merely for one specific group. There is a growing recognition on the part of employers that the perceived value of health insurance among employees is high relative to the actual cost, which may be actually quite small. It is therefore a very useful way of improving employer/employee relations. Moreover, health insurance is likely to become a more common part of recruitment packages designed to attract and retain skilled staff.

The economic advantages of worksite health promotion programmes experienced in North America are thus becoming increasingly relevant to the UK. Health insurance premiums here are likely to rise, as they have in the US, in parallel with medical costs in general, and organisations will be looking for ways of reducing their exposure in this respect.

Reduction in health insurance claims is one important aspect of the positive economic impact which worksite health promotion programmes can have. However, we should not forget that financial advantage can be achieved from the improvements in productivity and reduced absenteeism reported in many of the studies mentioned above.

Extrapolation from the American to the British experience so far as effects on productivity and absenteeism are concerned seems very reasonable, and there is no good reason to suppose that British companies could not achieve similar results. Whether improvements in productivity and reduced rates of absenteeism would, in themselves, cover the costs of a comprehensive worksite programme has yet to be determined.

'Proof' or probability

There is no absolutely conclusive evidence of the economic benefits of worksite health promotion programmes in the USA and certainly not in the UK. Indeed, it is questionable whether a definitive study could ever be performed, given the complexity of many of the issues involved and the cost implications. Even if the perfect study could be done for an individual company, the cynics would argue (with some justification) that the evidence obtained was only applicable to the particular company involved. It would be difficult to claim that similar programmes would bring similar benefits to different kinds of organisation and workforce.

However, the fact that 'proof' in a strictly scientific sense will probably never be available, should not concern us unduly. There is an enormous amount of uncertainty in medicine, both preventive and acute. There is even more uncertainty in economics. Indeed all scientific progress has been based upon uncertainty. This lack of guarantee should not be construed as a reason for inaction and apathy. This is manifestly not the way forward. The problems have to be attacked, not merely accepted. Our lives are not based upon certainties but upon probabilities.

In the cholera epidemic of 1854, the London doctor John Snow came to the conclusion that the variations in the incidence of disease throughout the city were due to polluted drinking water and not to 'bad air' as some had claimed. Of course Snow had no idea what was being carried in the drinking water and had no way of knowing, as we now do, that the disease was actually due to bacteria contaminating the water supply. Snow reached his conclusion from a statistical analysis of cases in various parts of the city. Having the courage of his convictions, he was able to demonstrate their accuracy by removing the handle of one Soho pump, thus compelling the local inhabitants to obtain their water from another, safer source. It was not until 1883 that Robert Koch discovered the organism which causes cholera.

Even the relationship between cigarette smoking and lung cancer is not actually proven. To be able to prove the association we would need to have an initimate knowledge of the mechanism by which smoking produces lung cancer, and this evidence is still lacking. However, although we cannot prove that smoking causes lung cancer, we do know that the relationship between them is so close as to be, for all practical purposes, *causal*. Thus the balance of probability is overwhelmingly in favour of the generally accepted idea that smoking causes lung cancer. And if we wait until absolute proof is available, we know beyond any reasonable doubt, that a great many individuals will suffer and die as a consequence.

The point is that if we keep on believing that proof is a prerequisite for action, we shall never do anything. All of our decisions are based upon probability, and the decision as to whether we should act or not must be based upon what the burden of the evidence tells us. Those of us who are trying to improve the health of the population by means of preventive medicine, are all too conscious of the limitations of our knowledge. But the fact that we cannot yet tell the whole story must not prevent us from acting upon what we already know.

So far as worksite health promotion is concerned, we already know a great deal. We know beyond any reasonable doubt that strategies designed to reduce risk factors in the workforce via scientifically based programmes of prevention are likely to have a significant beneficial impact upon the health of the whole workforce. As we have seen, the evidence supporting the claim that such strategies will result in significant economic benefit is much less secure. But recent studies of high calibre have yielded positive and very encouraging results. It is important for us to move forward by building upon these initial successes and basing our conclusions upon the evidence available. A healthy scepticism is appropriate, provided this does not become an excuse for inactivity.

Summary

In selecting the studies which I have presented in this chapter, I wanted to address two fundamental questions:

1. Does the available evidence suggest that worksite health promotion programmes bring about a significant improvement in the health of the workforce with a consequent reduction in the risk of premature disability and death?
2. Does the evidence suggest that worksite health promotion programmes can bring about significant economic benefits in terms of reduced health insurance costs, reduced absenteeism, reduced staff turnover and improved productivity?

The answer to the first question is an unequivocal *Yes*. The answer to the second question is a more cautious *Yes*. There is a good deal more scientific evidence to support the first proposition, than there is to support the second. However, I would like to emphasise that although the case for economic benefit is less compelling, those good quality studies that do exist have tended to support the proposition, rather than to refute it. This evidence is clearly of great importance and provides the foundation for future research. In five or ten years from now, we may be in a position to give a more definite *Yes* to the second question. At the moment the jury is still out.

CHAPTER 3

THE NEW DISEASES – THE SCALE OF THE PROBLEM

'It seems to me most strange that men should fear; seeing that death, a necessary end, will come when it will come'

Julius Caesar

In Western industrialised society we have replaced diseases such as tuberculosis, typhoid and other illnesses commonly associated with poverty, with a new group of diseases which are by contrast associated with affluence and progress. Heart disease, cancer, alcoholism – the list is a long one – are often regarded as an inevitable accompaniment to our growing development and prosperity. The price we pay is indeed a heavy one, but as we shall see later there is actually nothing inevitable about it. We have all heard people say 'well you have to die of something and a quick heart attack is probably as good a way as any to go'. Implicit in this sort of comment, whether conscious or unconscious, is the assumption that this quick heart attack will happen at some distant point in the future. It is natural and appropriate that we should think of ourselves as dying in old age and natural also that we should think of ourselves as being in good health and remaining active up until that point. In this sense, as Caesar says, death is a 'necessary end'. We all share Sir Richard Doll's ambition to 'die young as late as possible'.

But what is so alarming about the pattern of disease in our society today is that so many people are disabled or killed while they are still young, active and productive. It is the prematurity of their death or disablement which is so devastating and which brings incalculable grief and misery to so many. Because many of these individuals are in the most productive periods of their lives, the economic consequences both for the companies they work for and for the country as a whole are enormous.

In this section of the book I aim to provide some basic background information concerning the scale of our current health crisis and the burden, both medical and economic, which it imposes on all of us. These are, of course, only statistics but it is important for us to bear in mind that we are referring to people and not just numbers. In this and in the later sections of the book I have chosen to focus on five disease groups which collectively constitute the major part of our current morbidity and mortality. This is not to imply that there are not other important disease groups, but simply an acknowledgement of the fact that it is not possible within the scope of this book to give a comprehensive account of them all. The five disease groups are:

- Cardiovascular disease (principally heart disease and stroke)
- Cancer
- Stress and mental illness
- Alcoholism and other forms of drug abuse
- AIDS (Acquired Immune Deficiency Syndrome).

Cardiovascular diseases

Cardiovascular disease is a general term used to describe disease of the heart and arteries. It can be divided into two main groups: coronary heart disease and cerebrovascular disease. Coronary heart disease (CHD) affects the blood vessels supplying the heart muscle and leads to heart attacks. Cerebrovascular disease, on the other hand, primarily affects the blood vessels supplying blood to the brain and gives rise to strokes. (See Figure 3.1.)

Cardiovascular disease remains the most serious contemporary healthcare problem in Western industrialised society. In England and Wales in 1985 it accounted for the deaths of more than 270,000 people with heart disease and strokes accounting for approximately 75 and 25 per cent respectively.

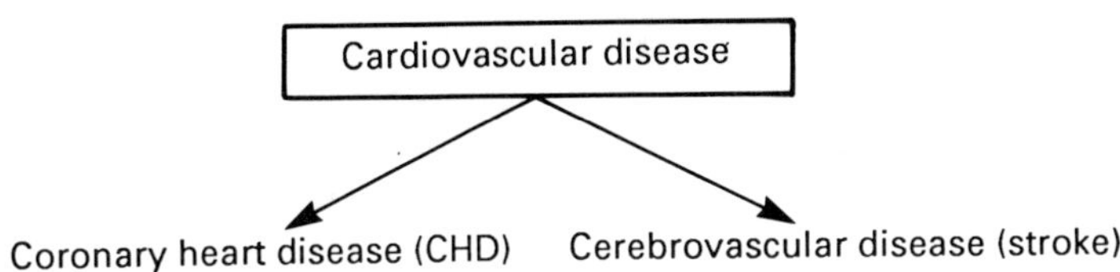

Fig. 3.1 *Cardiovascular diseases*

To set these figures in a broader perspective, we can say that disease was responsible for about half the total number of deaths recorded in England and Wales in 1985. (See Figure 3.2.) A similar pattern has long existed in the USA where cardiovascular disease accounts for almost one million deaths annually, more than all other diseases combined. About half of all cardiovascular deaths are due to coronary heart disease. It is estimated that approximately 16 million people in the USA have partial or severe disability as a result of cardiovascular disease, and about half of this total is probably due to coronary heart disease.

We shall now consider coronary heart disease and strokes separately.

Coronary heart disease (CHD)

This is undoubtedly the disease of the decade, the modern equivalent of the plague, although it does, of course, kill far more people than the plague did. Coronary heart disease is almost exclusively a twentieth-century phenomenon. Even in the 1920s and 1930s, the disease was regarded as an uncommon condition. In his classic *Textbook of Medicine,* the physician Sir William Osler wrote in 1920: 'Angina Pectoris is a rare disease in hospitals, a case a year is about the average, even in the large metropolitan hospitals.' And in

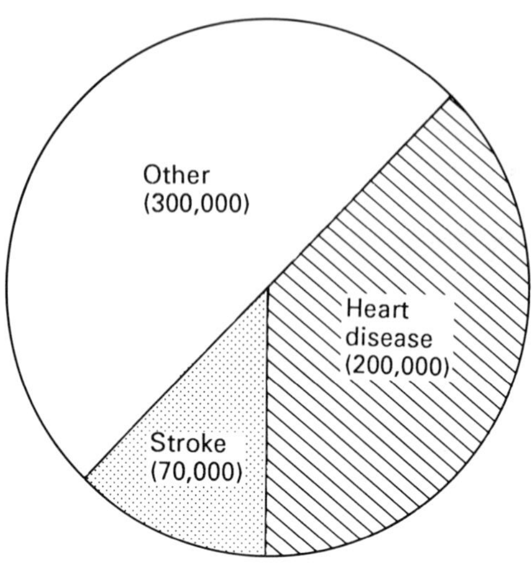

Fig. 3.2 *Deaths in England and Wales (1985)*

1928 there was a professional outcry when a question on coronary heart disease was included in an examination for doctors; it was felt that doctors were extremely unlikely to see such an obscure disease in practice.

Present day costs of this disease in both human and economic terms are staggering. Of the 200,000 or so deaths from heart disease in England and Wales in 1985, about 160,000 were due to CHD. Here are some facts about coronary heart disease which will help to set the problem in a broad context.

- One man in eleven can now expect to die of CHD before the age of 65.
- CHD will account for more than two out of five deaths each year in males aged between 45 and 64 years.

- In England and Wales CHD will kill one person every three to four minutes (the equivalent of more than one jumbo jet crashing every day of the year).
- CHD caused the year of one person *under 65* every 16 minutes throughout 1985.
- In 1985 the CHD death toll among persons aged 15 to 64 years, generated a loss of potential working life estimated at nearly 240,000 years, with males accounting for 82 per cent of the total.
- The cost to the nation of CHD alone, is estimated to be in excess of £1 billion per year.
- CHD is estimated to cost the NHS about £400 million each year, or the equivalent of £45,000 per hour for every hour of every day throughout the year.

While it is true that CHD tends to be much more prevalent in men than in women (about four times), we should not underestimate the significance of CHD as a cause of premature death among females. In 1985 the disease accounted for about 15 per cent of deaths of women aged less than 65 years, and about 21 per cent of those between 55 and 65 years. It is also extremely important to remember that CHD is the biggest *overall* killer in women.

In other countries too the problem is enormous. In the USA, for example, despite major progress during the past 15 years or so, CHD still accounts for more than half a million deaths a year, 200,000 of which occur before the age of 65.

In the statistics above we are dealing with death rates. But the problem of CHD is very much bigger than this because a great many people suffer heart attacks and survive or else have symptoms such as angina or breathlessness which are due to CHD. An even greater number, which it is obviously difficult to estimate accurately, have established CHD but have no symptoms and are therefore not aware of it. Half a million cases of CHD are diagnosed in Britain annually and findings of the British Regional Heart Study suggest that about 1.36 million males aged between 40 and 59 years in

England and Wales have some evidence of CHD; about one in every five of these (300,000 men) can be regarded as being severely affected by the disease. In advanced cases, CHD can lead to heart attacks and in males aged between 40 to 59 years in England and Wales it is estimated that there are probably as many as 35,000 heart attacks each year. In the United States it has been estimated that approximately 1.25 million acute heart attacks occur each year.

International comparisons are instructive, and Figure 3.3 gives the international league table for deaths due to CHD. You will observe that the countries of the United Kingdom are in the top 25 per cent of the league, with Northern Ireland having the highest mortality rate of all. The case of the USA is particularly interesting because ten or 15 years ago it would have been almost at the top of the league table. However, during this period, it has experienced a 30 per cent reduction in death rates from CHD, a dramatic trend which appears to have followed an increasing awareness among Americans of CHD risk factors. Canada, Australia, New Zealand and Belgium have also shown substantial improvement during the past decade or so.

The fact that CHD is primarily a disease of the Western, prosperous industrialised societies is graphically illustrated by the work of Dr Dennis Burkett. As a surgeon in a teaching hospital in East Africa for more than 20 years, Dr Burkett accumulated an enormous amount of data from post mortem examinations performed on Africans who had died at the hospital from a variety of causes. He found only one case of coronary heart disease each year. The rarity of the condition in the third world and underdeveloped countries is powerful evidence to suggest that the problem relates to our lifestyle, including smoking, diet, stress and lack of exercise, and possibly to environmental pollution.

Stroke

A stroke, as we shall see later is due to a degenerative disease

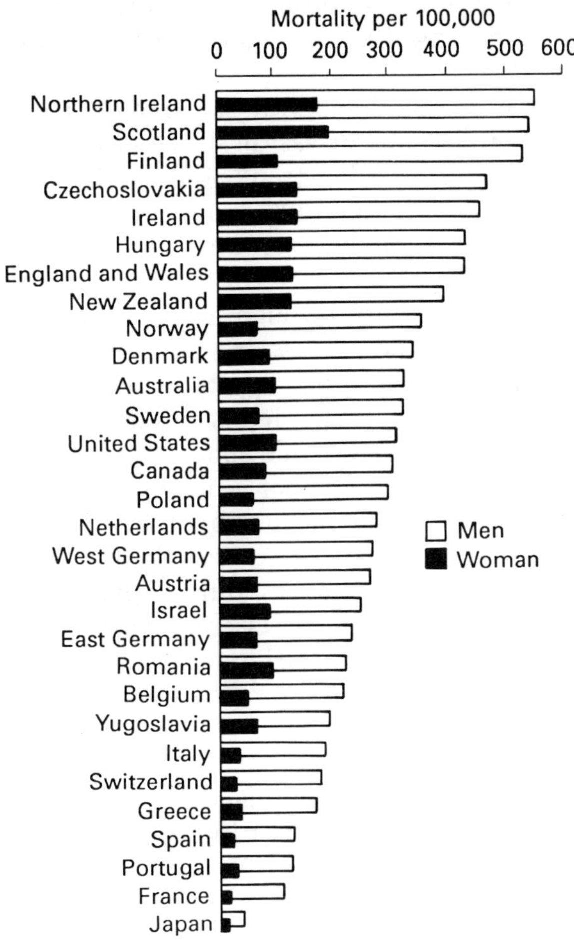

With permission of the Cardiovascular Epidemiology Unit, Dundee

Fig. 3.3 *Coronary heart disease mortality in men and women ages 40–69*

affecting the arteries of the brain. It is, in fact, identical to the disease process which occurs in the coronary arteries and which may lead to heart attacks. A stroke usually results in death or in serious disability, with some degree of residual paralysis affecting either the righthand or the lefthand side of the body.

In 1986 in England and Wales, about 70,000 deaths were due to cerebrovascular disease or stroke – roughly one quarter of all deaths due to cardiovascular disease. Each year approximately two people per 1000 experience a first stroke and about two thirds of these require some form of medical intervention.

Loss of productivity through absence from work is the conventional method of assessing the economic burden of a specific disease. In 1982 cerebrovascular disease was estimated to have accounted for 6.8 million days of incapacity, which represented about 1.8 per cent of the total absenteeism recorded for that year. In 1985 deaths of men and women under 65 meant about 60,000 lost years of working life. This amounted to less than 4 per cent of lost working life from all causes during that year, but it is certainly not insignificant. Cerebrovascular disease is estimated to have cost the NHS nearly £550 million in 1985 in England and Wales – approximately 4 per cent of total NHS expenditure.

It is important to point out that high blood pressure (hypertension), which is perhaps the major risk factor for cerebrovascular disease, has a high prevalence in the UK workforce. The World Health Organisation estimates that between 10 and 20 per cent of the population are hypertensive, but only half of these cases are actually known, and only half of those which are known are being treated. By 1991 there will be 350,000 more people aged 75 and over than there were in 1985 and cerebrovascular disease will therefore continue to make very significant demands upon our resources. High blood pressure is also a major risk factor for heart disease, so the health and economic implications of these estimates are very significant. As a potentially modifiable risk factor, hypertension is of major importance.

Cancer

The first thing to make clear about cancer is that it is quite wrong to think of it as defining a single disease; it is, rather, a whole collection – perhaps many hundreds – of different diseases. Whether the different forms of cancer have a single origin or a fundamental causal mechanism in common is not yet known.

Cancer is second only to heart disease as a killer in the developed world, and death due to certain forms of cancer, for example, lung, breast and skin cancer, is on the increase. Lung cancer accounted for 15 deaths per million of the population in 1918; by 1986 this figure had grown to 1036 per million. In the United Kingdom as a whole during 1984, nearly a quarter of a million new cancer patients were registered. On the basis of current trends about one in three of us will develop cancer at some stage during our life. In 1986 about 160,000 people died in the UK from various forms of cancer, and statistics suggest that about one in five of us will die from cancer of one form or another. Indeed cancer is now responsible for one quarter of all deaths each year.

A few dominant forms of the disease account for more than half of all cancer deaths – lung cancer causes 25 per cent of all cancer deaths (6 per cent of all deaths); colon and rectal cancer 12 per cent; breast cancer 10 per cent; and stomach cancer 7 per cent. (See Figure 3.4.)

Fatal cancers are more common in males than in females. In 1986, 83,000 men and 77,000 women died from cancer. Mortality rates for each sex and type of cancer, are shown in Figure 3.5.

The figures above indicate the scale of the problem but there is compelling evidence to suggest that a great deal of this terrible loss of life – perhaps even the majority – is preventable. Philippe Shubik of the US National Cancer Advisory Board has said: 'It is a universal opinion that cancer can be attributed to environmental factors in the main. I shall not belabour the matter of percentages as this seems to serve no useful purpose. Cancer is largely a preventable disease.'

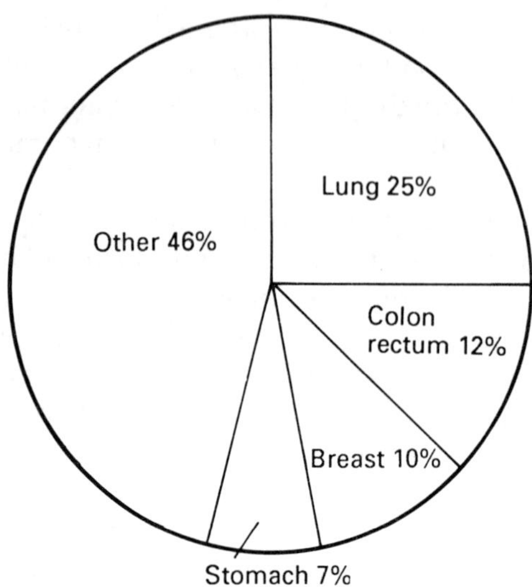

With permission of the Cancer Research Campaign

Fig. 3.4 *Cancer mortality in the UK (1986)*

Stress and mental illness

Mental illness which is easily the most prevalent of our contemporary health problems. Indeed, about one man in twelve and one woman in eight can expect to enter a psychiatric hospital at some time in their life, and at the present time approximately half the NHS beds are taken up by patients with mental illness of one form or another. In addition to this, there are thousands of individuals with less serious forms of mental disturbance who are being treated on an outpatient basis, usually, although not necessarily by their family doctor. The much-debated difficulties of defining 'mental illness' make the size of the problem and its economic consequences impossible to quantify. However, we do know that there has been a very rapid rise in the

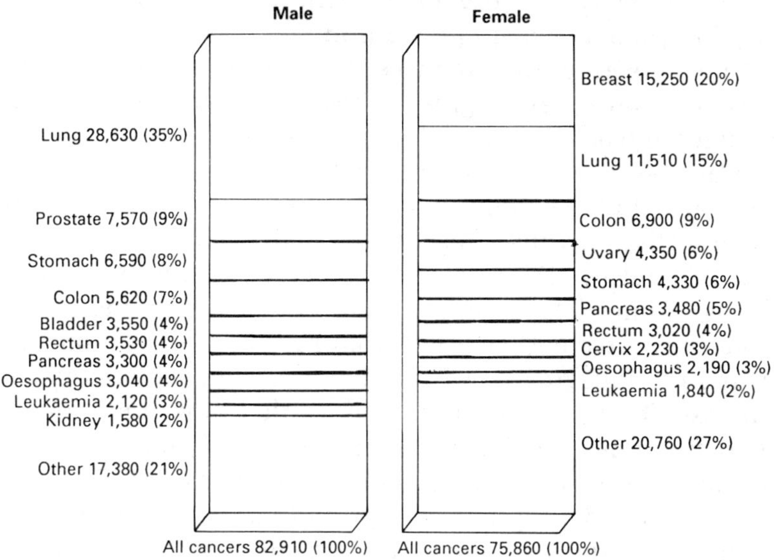

With permission of the Cancer Research Campaign

Fig. 3.5 *Number of cancer deaths in the UK (1986)*

number of people seeking psychiatric help and in the number of tranquilliser prescriptions being issued – in the UK in 1983 more than 21 million prescriptions; in the USA in 1980, nearly 35 million. In Britain, on average, three quarters of a million people take tranquillisers of one form or another every single day. Indeed, a report by the Office of Health Economics in 1983 called Britain 'a highly neurotic nation', due to the high consumption of tranquillisers, sedatives and antidepressants, compared to France, Italy and Spain. This problem could rightly be termed an epidemic.

It is important at this point to exclude from our discussion the types of mental illness described as psychotic – serious disturbances of mental function such as schizophrenia and depression – and also those disorders which are primarily due to structural disease of the brain – dementia and mental handicap. We exclude these conditions because they are not amenable to any form of resolution in the working environment. We know so little about them that we cannot yet do anything about primary prevention.

The less serious disturbances of mental function, however, may well have their origins in lifestyle and environment, of which the workplace is a significant part. Nowadays, extraordinary emphasis is placed upon the role of stress in causing ill-health, but it is important to realise that not all forms of mental illness which could be related to environment, are caused by stress. Depression can result equally from a lack of stress, from a lack of stimulation, from the inevitable sameness and boredom of everyday routine. The relationship of an individual to his environment is a complex, interactive process. It is quite wrong to believe that those individuals who are under stress or pressure are the only ones likely to become depressed or show symptoms of anxiety. To focus too much on identifying those individuals who are subjected to outward signs of stress, is to neglect an equally large number of individuals, the origins of whose illness is completely different.

Nevertheless, stress appears to be an important and growing contributory factor in many forms of illness, and the growing recognition of this fact is perhaps one of the main reasons why so many companies and organisations seem to be so preoccupied with it. As we shall see later, stress is a term which is not actually amenable to precise definition, and therefore its economic consequences can only really be guessed at. Current estimates suggest that it may be accounting for as many as 30 million lost working days each year, and one university study estimated that cost between 1 and 3 per cent of Britain's gross national product. Other estimates have suggested that stress is currently costing British industry between £3.5 and 10.5 billion a year. Stress and its consequences are discussed in much more detail in Section 2.

Alcoholism and other forms of drug abuse

Alcohol abuse and addiction are, without question, extremely serious health issues in Britain today. It is especially worrying because, despite increasing evidence of the vast scale of the problem, the message that something urgent needs to be done, continues to fall on deaf ears. Of course, the problem is hardly a new one. In 1726 the Royal College of Physicians made a submission to the House of Commons about what was termed 'a great and growing evil', and this phrase became the title for the most recent report from the Royal College of Physicians on Alcohol Abuse, published in 1987.

Alcohol is a poison, which can exert a direct toxic effect on the brain, liver, heart – indeed almost any organ or system in the body. In the UK during the past 30 years or so the consumption of beer, wines and spirits has risen sharply. The total number of deaths from cirrhosis of the liver is one of the most useful indicators of the trends in alcohol-related disease and this tends to rise and fall in parallel with alcohol consumption. A particularly worrying feature of these trends is the growing problem of alcohol abuse and addiction in young people.

Hear are some basic facts about the alcohol problem in this country:

- Evidence suggests that there may be as many as three million excessive drinkers in England and Wales. Estimates of the number of problem drinkers vary between 0.5 and 1.3 million and of this total number somewhere between 70,000 and 240,000 can be regarded as having severe alcohol addiction.
- From the above, it seems that about one in ten of the adult population may have some alcohol-related problem.
- About one in five of all men admitted to medical wards has a problem related to alcohol abuse.

- The British spend more on alcohol than they do on clothes, cars, hospitals, schools, or universities. In 1981 this came to over £11 billion, representing 7.5 per cent of total consumer expenditure.
- In 1983/84 the government collected almost £6 billion in excise duty and value added tax from the sale of alcohol.
- Exports of alcohol are worth more than £1 billion each year.
- At least 750,000 people are engaged in producing and selling alcohol.
- Between eight and 14 million working days are lost each year as a result of sickness absence due to hangovers.
- A report published by the Royal College of Practitioners in 1986 estimated that 40,000 deaths may be caused annually in Britain by alcohol. A 1988 study published in the *British Medical Journal,* gave an estimate of 28,000 deaths each year associated with alcohol abuse in England and Wales.
- Alcohol abuse generates considerable cost. Lost output in industry in England and Wales from alcohol consumption was estimated at £1.3 billion in 1985. This was broken down as follows;

Sickness absence	£641.5 m
Unemployment	£144.7 m
Premature death	£567.7 m

- In the United States the annual cost of alcohol abuse may be in excess of $113 billion.

It is useful for the purpose of this discussion to divide the effects of alcohol into two groups: acute and chronic. (See Figure 3.6.) There is clearly considerable overlap between the acute and chronic effects. This is because many people who have acute problems go on to develop chronic problems, and chronic sufferers occasionally have specific problems

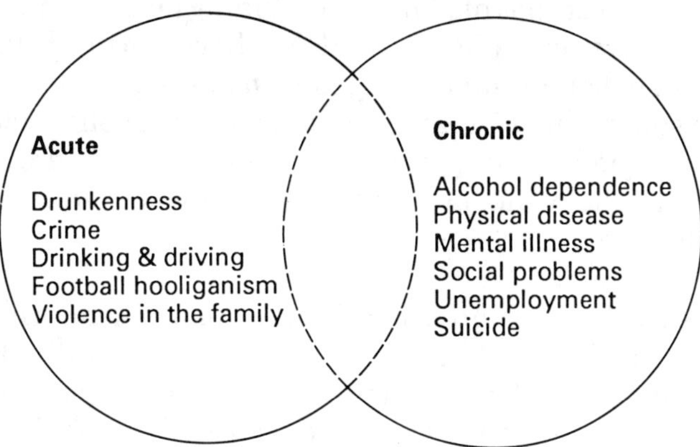

Fig. 3.6 *Acute and chronic effects of alcohol*

related to acute intoxication. We shall now consider each of these effects in a little more detail.

Acute effects

1. *Drunkenness.* In 1983 there were approximately 110,000 convictions for drunkenness in England and Wales; the overwhelming majority involved males (100,000). In Scotland in 1981 there were approximately 12,000 convictions for drunkenness.

2. *Drinking and driving.* In 1982 there were almost 64,000 convictions for drinking and driving offences in England and Wales, and again the overwhelming majority of these involved males (61,000). In Scotland in 1981 there were just over 11,000 convictions for drinking and driving. However, a Home Office study suggests that only one in 250 episodes of drunken driving results in a conviction. This is likely to be a conservative estimate.

3. *Crime.* The abuse of alcohol is also strongly associated

with crimes against persons and property. Alcohol is a major factor in many crimes of violence including murder and rape, and offenders are often found to have been drinking immediately before committing the offence. It is well known that prisoners have a high incidence of alcohol-related problems. Football supporters represent a specific group of people in whom alcohol abuse and violence are clearly associated.

4. *Violence in the family.* The families of problem drinkers probably suffer at least as much as the individuals themselves. In the United States as many as 80 per cent of cases of family violence are associated with alcohol and a similar figure probably applies in this country. Conflict between parents is likely to turn against children who themselves become victims of violence associated with alcohol.

Chronic effects

1. *Physical disease.* The most important thing to recognise is that a great deal of alcohol-related illness and even death occurs in people who do not think of themselves and may not be considered by their doctors as either alcohol abusers or problem drinkers or in any way dependent on alcohol. Evidence suggests that death from cirrhosis of the liver, accidents, suicide and other causes associated with alcohol are grossly underestimated. To some extent this may be because doctors signing death certificates are rather unwilling to specify alcohol abuse as the sole cause of death or as a contributor factor. Sometimes they are completely unaware that alcohol played any part at all.

Alcohol can cause serious damage not only to the liver, but to the brain, to the heart, to the kidneys and to almost every organ or system in the body. The Royal College of General Practitioners has estimated that about 40,000 deaths a year may be caused by alcohol and as many as 500 young people may die each year while drunk, representing 10 per cent of all deaths under 25. An estimate of the number of life years lost through alcohol abuse in England and Wales in 1982, gave a low figure of 119,000 and a high figure of 192,000. At least one third of all deaths in traffic accidents are associated with alcohol (about 1700 deaths per year).

In Britain and the United States a fifth of all acute male admissions to medical wards are associated with alcohol; between 10 and 40 per cent of patients in general hospital beds are problem drinkers and up to one quarter of all cases of self-poisoning admitted to casualty departments may be associated with alcohol.

Finally, because alcoholics tend to neglect themselves and their families, malnutrition may also be a feature of chronic alcohol abuse.

2. *Psychological damage.* An extensive report on the psychological effects of alcohol was published by the Royal College of Psychiatrists in 1986. An enormous variety of mental illness may be associated with alcohol abuse. Depression, anxiety, personality disorders and paranoid states can all be associated with heavy alcohol consumption, but there may also be serious structural damage to the brain itself in some cases. In 1981 there were almost 18,000 alcohol-related admissions to mental hospitals in England and Wales – about 12,000 men and 6000 women.

Suicide is between 20 and 60 times more common among excessive drinkers than in the general population. Evidence shows that anything up to 80 per cent of those who kill themselves have been drinking, and that a suicide attempt is much more likely to be successful if the individual has been drinking.

3. *Social problems.* In Britain there may be more than 100,000 homeless people (mostly men), and a great many of

these have severe alcohol problems. Alcohol abuse leads to family problems and marital breakdown. At least one third of problem drinkers list marital discord as one of their most serious problems and one third of divorce petitions cite alcohol as a contributory factor. As well as suffering violence from drunken parents, children may also be affected by family break-up, poverty and debt resulting from alcohol abuse.

As well as creating difficulties with family and friends alcohol problems can have a serious effect on employment. Employees with alcohol problems have an increased likelihood of accidents and a very much higher level of absenteeism. Drunkenness at work, excessive absenteeism and difficulties with working relationships can very often precipitate an individual's dismissal. This in turn may exacerbate an already difficult financial situation, leading to more discord in the family, further violence and more drunkenness; and so the vicious circle continues.

Alcohol in the workplace

Although problem drinkers are more likely to have accidents at work, take more than average sick leave, and produce a below average work performance, the fact is that 75 per cent of such people are in full-time employment. With the total number of days lost to industry each year through alcohol-related problems estimated to cost about £650 million a year, it is clear that problem drinkers are taking their toll. Several studies have demonstrated the likelihood of heavy drinkers being involved in accidents at work. It is known that anyone who consumes three pints of beer or its equivalent (six units) before driving, is up to six times more likely to have an accident. There is no reason to suppose that the same dangers do not exist in the factory. Alcohol affects judgement and decision making, and lack of leadership and indecision may be signs of a problem in senior management.

The multitude of problems raised by excessive drinking cannot be separated from each other. The physical, psychological, social, emotional and economic consequences are all interrelated and interdependent. Scarcely any part of the problem drinker's life remains unaffected, and for many the circle of excessive drinking, violence, problems at work, problems at home, increasing financial difficulties, leading to more drinking becomes a downward spiral of despair. The effects of alcohol in terms of decreased productivity, accidents, crime, physical and psychological problems and disruption of family and social life are enormous.

In the last few years an increasing number of employers have developed policies and programmes designed to help the problem drinker rather than to dismiss the person or to ignore the problem altogether. The scale of the problem and the knowledge that it may be affecting important and skilled personnel are compelling financial reasons for corporate organisations to come to grips with it. In Chapter 11 we look at what strategies are available and how they can be effectively implemented in the workplace.

Drug abuse

Of course, alcohol is only one form of drug abuse although there are good reasons for considering it separately. Obtaining evidence relating to other forms of drug abuse is much more difficult, because the drugs are largely illegal and users are therefore more reluctant to divulge information.

In America the abuse of drugs has become a truly awesome problem, affecting all levels of society from the very poor to the fabulously rich. A comprehensive study carried out by the National Institute on Drug Abuse in 1985, revealed just how pervasive the problem has become. It showed that almost 62 million Americans over 12 years of age – out of a total of 190 million – had at sometime used marijuana or hashish, and 18 million continued to use it. More than 22 million had used cocaine and almost six million were using it at the time of the survey. About two million had used heroin, about three quarters of them over 26 years of age.

Some encouraging trends in drug use have emerged in recent years. Since 1985 an annual survey of high school seniors has shown a decline in the use of almost all drugs by 18-year-olds. In 1988 about a third of high school seniors reported some use of marijuana, compared with the peak in 1979 when half said they had used it. Cocaine use dropped from 13 per cent in 1986 to 8 per cent in 1988. Heroin use in this group has, however, remained constant for almost a decade at about 0.5 per cent.

The use of cocaine by the American middle class is declining but, at the same time other drug problems are growing. 'Crack' (a cheaper smokable and highly addictive form of cocaine) came on the scene in 1985, and this has increased the number of addicts considerably, especially among the poor and the unemployed. It has also provoked an epidemic of drug-related murders, particularly in Washington. Crack tends to produce dependency on other drugs, and has produced a revival of heroin addiction in the northeastern states of America. New York city with an estimated 260,000 heroin addicts, has about half the regular heroin users in America. The use of newer substances such as 'crank', a methamphetamine stimulant, is likely to produce a further surge in drug abuse and addiction in the 1990s.

While recent studies have shown an overall decline in the use of drugs, the problem is still an enormous one, not only in terms of human suffering, but also economically. Accurate estimates are difficult, but one study in 1984 suggested that the annual cost to the US economy in terms of lost productivity, was in excess of $25 *billion*.

Nobody knows the full extent of the drug abuse problem in the UK, but there is good evidence that it has risen dramatically in the past decade. The number of 'addicts' notified to the Home Office doubled between 1983 and 1985 and almost doubled again the following year. At the moment it is in excess of 9000, but this represents only the tip of the iceberg. Some authorities estimate that it could be ten times higher or even more. A survey carried out by the Institute of Psychiatry in 1986, revealed that GPs were seeing 40,000 users each year. Not all of these drug users are on 'skid row'. Indeed, evidence shows that more than 25 per cent of clients applying for help at clinics and drug centres are in regular employment. Other forms of 'substance' abuse are also important. Everyday domestic items such as glue, hair-spray, lighter fuel, cleaning fluids and other solvents, are cheap and readily available over the counter. Not surprisingly, the 'glue-sniffing' phenomenon is particularly common in younger people in whom it may be both addictive and lethal.

We should not forget that there are drugs which are legally prescribed by doctors, but which nevertheless result in dependency and impaired efficiency for many people. Evidence suggests that as many as three million people may be dependent upon tranquillisers and this has obvious implications for industry.

Since we can only guess at the real scale of the drug problem, meaningful estimate of its potential cost to industry is not possible. It is probably significantly less than the cost of alcohol abuse, but since we know that it is increasing rapidly, there is absolutely no justification for complacency.

Acquired immune deficiency syndrome – AIDS

Ever since the disease now known as AIDS was first reported in 1981, two questions have preoccupied us. How far and how fast is this disease, which is almost invariably fatal, likely to spread? And does it threaten the vast majority of the population who are not in high risk groups such as homosexuals, bisexuals or intravenous drug users?

The answer to the first question is that AIDS has already

spread to virtually every country in the world and the numbers of cases and deaths continue to rise. What the eventual end point will be is not known. Any predictions are subject to considerable uncertainty, partly because of the natural biological variability of the virus and the population in which it is active, and partly because of the limitations of data collection in many of the countries in the world.

The answer to the second question is that AIDS threatens all of us. It is still not certain whether we will see the explosion of AIDS in the heterosexual population which some people have predicted. What is clear, is that the way to deal with this epidemic is by taking preventive action. If it is possible to produce a vaccine – and there are many who doubt that it is – it is still many years away. None of the treatments currently available is likely to have any major impact on the spread of the disease in the community. At best such treatments can only relieve some of the symptoms and perhaps prolong life.

We shall be discussing the various manifestations of this disease and in particular the implications of AIDS for employers in Chapter 14. The following are some basic figures concerning the progress of the disease to date, in Great Britain and in the United States.

The United Kingdom

By the end of March 1989, a total of 2192 people had been reported as having AIDS, of whom 1149 had died. At the end of December 1988 the comparible figures were 1982 and 1059. Males accounted for the overwhelming majority (96 per cent) of all these AIDS cases.

Most cases (82 per cent) occurred in homosexual/bisexual males. In 81 cases (3.5 per cent of the total), intravenous drug abuse was thought to be the sole mode of infection or a major contributory factor. Heterosexual transmission was thought to have occurred in 90 cases (4 per cent).

It is not known how many individuals currently carry the AIDS virus without having yet developed the full blown syndrome. The number in England and Wales has been estimated at between 20,000 and 50,000, of whom 13,000 to 30,000 are homosexuals. However, many experts regard these figures as grossly underestimated, and believe that the more realistic figure is somewhere between 80,000 to 100,000.

The total number infected by heterosexual contact so far is estimated to be between 2000 and 5500. Unfortunately, the virus continues to be spread into the heterosexual section of the population by infected drug users, and a more realistic figure is probably between 6000 and 17,000.

An expert group, chaired by Oxford statistician Sir David Cox, was asked by the Ministry of Health to predict the numbers of HIV-infected individuals and AIDS cases over the next few years. While recognising the enormous difficulties inherent in such an exercise, the report suggests that by the end of 1992 in England and Wales, between 10,000 and 30,000 AIDS cases will have been diagnosed, with deaths between 5000 and 15,000. Given the fact that the vast majority of HIV individuals will eventually die from the disease, it is clear that the cumulative mortality is enormous.

A recent report from the Office of Health Economics (OHE) also draws particular attention to the potential economic effects. Assuming that the mean age of death from AIDS is 35 years, the loss of potential working life through AIDS fatalities in the 1990s could amount to as much as 165,000 years. This total substantially exceeds the losses for males from cancer (49,000 years), suicide (57,000 years), and even road accidents (97,000 years). Indeed, only coronary heart disease (195,000 years) causes more lost years of potential working life. The economic consequences of all this are formidable.

The United States of America

During the Vietnam war the number of Americans killed in battle was approximately 58,000. In 1989 the number of

Americans dying from the AIDS virus will exceed that number. By the end of 1991 the United States National Academy of Sciences predicts that there will be 270,000 cases and 179,000 deaths. In 1991 alone it is predicted that there may be 74,000 new cases and 54,000 deaths, costing between $8 and 16 billion annually in healthcare. At the moment it is estimated that about 1.5 million people in the United States have been infected by the AIDS virus and we know that the vast majority of these people will eventually contract the disease and die. Very few AIDS victims live for more than three years after the disease first becomes manifest.

In 1986 the Reagan administration committed $234,000 million to AIDS research and education combined. By 1990 the total cost of the research and education programmes will probably exceed $2 billion a year.

We have concentrated here on Great Britain and the USA, but, of course, AIDS is a truly global problem which affects every country in the world. Given the scale and the rate of growth of the disease, it is not inconceivable that the AIDS virus will become an agent of historical change, in the same way that the Black Death piled up 50 million corpses in Europe and ushered in the Renaissance. The African continent, in particular, is under the gravest threat. The total number of Africans carrying the disease is estimated by the World Health Organisation to be anything between 5 and 10 million. All are infectious and infecting others. But the tragedy is magnified by the fact that currently not a single African country has the resources, either financial or medical, to counteract the ravages of the virus. Even the USA, the richest country in the world, has been unable to stop the march of this disease. So what chance does a country like Uganda have which cannot even afford the bleach or disinfectant needed to clean down the hospital tables?

PART 2

CHAPTER 4

EXERCISING THE CORPORATION

'If some of the benefits accruing from regular exercise could be procured by any one medicine, then nothing in the world would be held in more esteem than that medicine.'

Francis Fuller, 1705

During the past decade or so we have witnessed a tremendous increase in interest and participation in many sporting and leisure activities, especially running and aerobics. There may be 20 million regular joggers in the USA and as many as 2 million in the UK. Marathon running too seems to have caught the public imagination and many thousands of hitherto sedentary middle-aged individuals are donning their tracksuits and running shoes, intent upon completing the mystical 26 miles 385 yards. The London Marathon has established itself as an extraordinarily popular event in the sporting calendar with more than a fourfold increase in the number of participants since the first race in 1981. (See Figure 4.1.)

But what are we to make of all this? What are the reasons for the emergence of this phenomenon, and is there any evidence that this increase in leisure activity is doing us any good? To the first question there would appear to be three main answers.

First, the increased availability of leisure time means that many people now have an opportunity to develop a sporting interest of one kind or another. Second, more and more people are realising that regular exercise may be of some benefit in terms of health, although they are often not sure *how* these benefits are brought about.

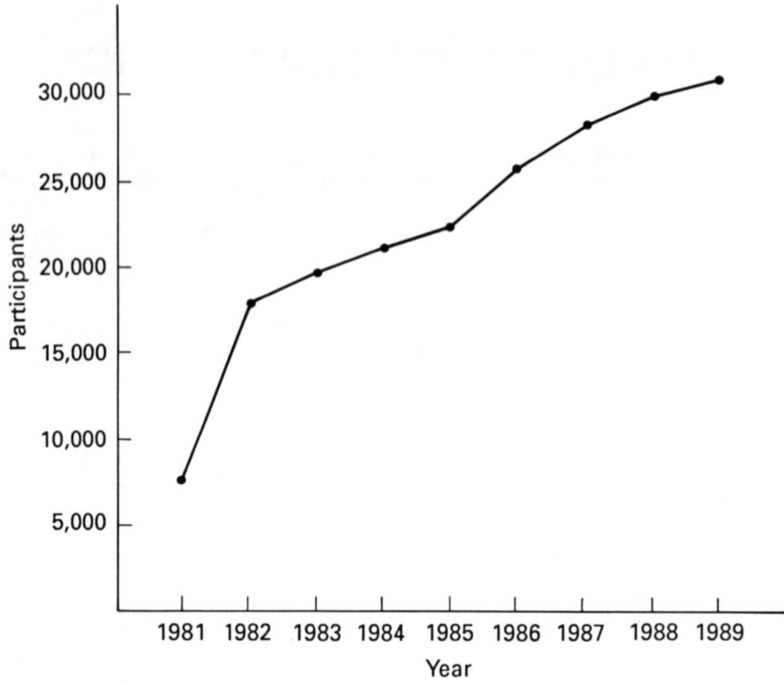

Fig. 4.1 *Participants in the London Marathon: 1981–1989*

The third important reason is that exercise has become fashionable and is heavily promoted by the media and advertising industry as a means to becoming slim and attractive. There has been no shortage of self-styled 'experts' keen to cash in on the commercial opportunities. Any number of film stars and TV personalities have become overnight authorities on nutrition, weight loss and exercise, each offering some 'unique' plan which will ensure health, beauty, success and longevity. Exercise, for many people, is therefore simply a means of improving personal appearance in response to growing social pressures, and any health benefits are a useful side effect. This is not necessarily a bad thing, but the obsession with improving personal appearance has tended to divert attention away from a far more important issue: the relationship between exercise, health and disease.

This chapter examines the role of exercise in disease

prevention, and suggests how its promotion within the workforce can be of substantial benefit to the corporation.

Promoting exercise

The growth in leisure activity has mainly been promoted through television, some sporting organisations and the Health Education Authority. Corporate organisations in the UK have, with some exceptions, done very little to promote the case for exercise among the workforce. Their involvement tends to be limited to the support of specific sporting events – for example, Kelloggs Tour of Britain Cycle Race – as a means of obtaining valuable publicity.

In the USA the attitude of corporate organisations is significantly different. Indeed the growth in interest in physical fitness among US companies has been staggering. Since the 1974 Canadian conference on employee physical fitness, corporate physical activity programmes have flourished, and today as many as 50,000 companies in the USA and up to 1000 companies in Canada are involved in some aspect of fitness promotion among the workforce. These companies include Rank Xerox, Century Insurance, Prudential Insurance, Johnson & Johnson, IBM and Kimberly Clark which has one of the most elaborate and sophisticated programmes, involving 15 full-time staff and a $2.5 million facility.

Of course, interest in exercise as a means of promoting a healthy workforce is not an exclusively American activity. Japan and many Scandinavian countries, in particular, have recognised the potential benefits and have been running successful programmes for a number of years.

Part of the problem in this country is that many health professionals have been slow to recognise the value of exercise both as a means of promoting overall health, and also as a specific form of therapy.

The case for exercise

Since the time of the ancient Greeks, physical exercise has tended to be associated with good health. Man has pursued recreational and competitive activities on the assumption that a fit individual is a healthy individual. As we shall see later, this is not necessarily true: it is possible to be in sound health but unfit; it is equally possible to be very fit but unhealthy.

Before the age of technology, many occupations demanded a lot of physical activity and energy. Today many more people have to rely on exercising during leisure time as a means of maintaining fitness. Unfortunately, most people simply do not bother. Despite the tremendous increase in interest, the fact is that physical exercise is still a minority activity. For every one person running the London Marathon there are 100 other people sitting at home watching him! Of course, no one would suggest that we should all become marathon runners – quite the contrary! But if we were to take all of the activity performed by the *minority* and spread it equally among the *majority* then the potential health benefits for the population as a whole would be enormous. In other words, it would be infinitely better to have all the population classified as moderately active rather than one tiny section of the population classified as highly active.

Since the early 1950s, scientists and physicians all over the world have studied the relationship between exercise and disease, particularly heart disease. Many of these studies have come to be regarded as classics in the now extensive literature on exercise. Some of the more important of these are discussed briefly below.

The London Transport study

In 1953 Professor Jerry Morris of the London School Of Hygiene and Tropical Medicine mounted a large-scale study on the London Transport bus drivers and bus conductors. He had been prompted to investigate this group of workers because of the observation that bus conductors tended to have fewer heart attacks than bus drivers. It was clear to Professor Morris that the essential difference between these two groups of workers was the amount of physical activity their jobs demanded. In those days bus conductors used to run up and down the stairs taking fares for most of their average working day.

About 31,000 men aged 35–64 were included in the study, and analysis of the data confirmed the original observation. Not only did the conductors have 50 per cent fewer heart attacks than the drivers, but drivers also suffered twice as many *fatal* heart attacks. Not only did the bus conductors have less heart disease than the drivers, but what disease they did have appeared later in life and was less severe.

Although this study was criticised on many accounts, it was nevertheless a landmark in our understanding of the relationship between exercise and heart disease. Morris and his colleagues had tested an important hypothesis and found the evidence to support it. During the next 30 years a great many large-scale and methodologically sound studies were carried out which confirmed the results from these initial observations and Morris himself contributed in no small measure to much of the research which was to follow.

The British civil servants study

More than 20 years after his pioneering study of the London Transport workers, Professor Morris published a further major survey in *The Lancet*. The title of the paper was 'Vigorous Exercise in Leisure Time and the Incidence of Coronary Heart Disease'.

In this survey Morris studied the relationship between leisure exercise and CHD in 17,000 civil servants from all over Britain. Participants were required to complete a questionnaire relating to weekend leisure activity patterns, and over a number of years this was compared with the number of fatal and non-fatal heart attacks. For those men engaging in vigorous leisure activity during the two sample days, the risk of fatal heart disease was only about 40 per cent that of their colleagues reporting no vigorous exercise, and for non-fatal heart attacks the risk was 50 per cent. In other words, not only did the vigorous exercisers have much less heart disease, they were also much less likely to die from it.

There are a number of important additional points. First, the fact that Morris opted to study leisure activity patterns is illustrative of his recognition that because most occupations in advanced societies have become more sedentary, so any contribution to public health is more likely to be derived from exercise taken in leisure time.

Second, Morris also demonstrated that the *intensity* of exercise, as well as the amount of exercise, was important. He defined 'vigorous' physical activity as a peak energy expenditure of 7.5 kilocalories per minute, equivalent to heavy industrial work. Examples of such activity patterns included jogging, cycling, swimming, tennis, brisk walking and hill climbing. The duration of physical exercise required to confer benefit appeared to be a total of one hour during the course of the weekend (that is two 30-minute sessions). Thus a threshold level for both the intensity and the duration of the exercise was defined.

Finally, and very importantly, the beneficial effect of this vigorous exercise was seen throughout the age range being studied (40–60 years), and especially in late middle age. In other words, age is not in itself a contraindication to exercise!

The San Francisco dock workers study

Ralph Paffenbarger and his colleagues studied the effects of physical exercise in San Francisco longshore men (dock workers). About 6300 subjects were divided into three groups – heavy, moderate and light – according to the amount of physical activity involved in their daily work. These subjects were followed for a period of 22 years (1951–1972), and the number of heart attacks was carefully documented. During the study period, 598 longshore men died from heart disease, and of this total number of deaths, 66 occurred in the heavy work group, 107 in the moderate and 425 in the light. Taking into account age-specific and age-adjusted death rates, there was an 80 per cent greater risk of fatal CHD developing in the sedentary workers than in the more vigorous longshore men. Moreover, this increased risk of heart disease in the sedentary group persisted even when allowance was made for the effects of other risk factors such as cigarette smoking and high blood pressure. In other words, exercise was shown to be an entirely *independent* risk factor.

Like Morris, Paffenbarger was able to demonstrate a minimum threshold for exercise intensity (5.2–7.5 kilocalories per minute), before the beneficial effects of physical activity were observed. Most of the benefit was found in the high activity group while the moderate and light activity groups showed no significant benefit.

Once again, these data supported the hypothesis that exercise protects against CHD.

The Harvard University alumni study

Paffenbarger and his colleagues subsequently examined the relationship between physical activity and heart disease in about 17,000 Harvard University male alumni (old boys), aged 35–74 years. The subjects in the study were classified according to the number of hours per week that they were involved in physical exercise and also by their energy output in kilocalories per week. Analysis of these data once again indicated that high level energy expenditure was protective against both fatal and non-fatal coronary disease. In fact, there was a 64 per cent greater risk of heart attacks in those with sedentary recreational pursuits than in those who took vigorous exercise. Moreover, the benefit existed at all age levels between 35 and 74 years.

Another extremely important point which emerged from this study, is that the beneficial effects of exercise cannot be stored for a long time. Those subjects who were physically active as students, but who became sedentary when they left university, showed no reduction in heart attack risk. Benefit was demonstrable only for those who *continued* a physically active lifestyle after having left university. In other words, it is not possible to put a lot of physical activity 'into the bank' in the early years and expect that this will provide benefit for the remainder of your life.

Paffenbarger and his colleagues have continued to follow this group of alumni, and a further report was published in the Journal of the American Medical Association in 1984. The most recent data confirm the original observations: sedentary alumni, even ex-athletes, have a high risk when compared with their physically active colleagues. Moreover, exercise continues to exert benefit *independent* of all other risk factors including smoking, obesity and hypertension. The authors conclude that 'the current exercise revolution may improve lifestyle, cardiovascular health and longevity'.

These examples are but a few of literally hundreds of studies examining the relationship between exercise and CHD. (A bibliography prepared by Professor Fentem and others at the Department of Physiology and Pharmacology in the University of Nottingham entitled 'Exercise and Health' contains no fewer than 1334 references.) The overwhelming majority, and this includes *all* the better quality studies, support the hypothesis that exercise produces a significant measure of protection against heart disease.

Exercise and ageing

The studies mentioned have focused on the beneficial

effects of exercise in terms of reducing the risk of cardiovascular disease. But of equal importance is the potential for regular exercise to preserve bodily function and to prolong *active* life expectancy, not just life expectancy.

In an extremely important paper in the *Annals Of Internal Medicine,* Drs Bruce and Larson of the University of Washington, reviewed the evidence that exercise can continue to provide major benefits well into old age. Exercise helps to maintain muscle strength and flexibility, preserves bone density, promotes psychological wellbeing and maintains cardiovascular efficiency and fitness. This means that instead of experiencing physical decline from the age of 50 onwards, as many people do nowadays, it is possible for us to remain active for a much longer period. Thus, even if we do not actually live any longer, we are 'old' for a much shorter period of time. The authors conclude that on the basis of current evidence, the appropriate use of exercise can add up to *ten years' active life expectancy* to the average middle-aged male.

The mechanisms

The precise mechanisms by which exercise reduces the risk of heart disease have not yet been fully explained, but some of the most important factors involved are discussed briefly below.

1. Improved cardiovascular efficiency

Regular exercise improves the efficiency of the heart muscle, by making it contract more forcefully and effectively. In an average, untrained person, the normal pulse rate is about 72 beats per minute. In a highly trained athlete, the pulse rate may be as low as 40 beats per minute. If the heart of the athlete can achieve as much in 40 contractions as the ordinary individual does in 72 contractions, then clearly the athlete's heart must be pumping more blood per contraction. Athletes do indeed have a substantially higher 'stroke volume' (the amount of blood ejected from the heart during each contraction) than non-athletes. Part of this can be explained by the fact that regular training increases the size of the heart, and the capacity of its chambers also increases by as much as 30 per cent. It is easy to see, therefore, that the heart of an active person is much more efficient in delivering blood and oxygen to the body's tissues.

2. Direct effects on the coronary arteries

The progressive clogging up of the coronary arteries with fatty substances is the essence of coronary heart disease; so anything which increases the diameter of these arteries and therefore improves coronary blood flow, is obviously beneficial. Post mortem studies on athletes have revealed main coronary arteries with diameters up to three times that of normal individuals. Whether this is an effect of exercise as such, or an innate characteristic is not proven. Studies on human beings are fraught with technical and ethical difficulties but there is good evidence from animal models that exercise not only increases the diameter of the coronary arteries but also substantially reduces the development of fatty deposits in the lining of the arteries.

Kramsch and his colleagues at the Boston University Medical Centre have done some fascinating work on the effects of exercise training on the coronary arteries of monkeys. The animals were randomly divided into three groups of nine animals each, then studied for a period of 36 months. One group was fed a high fat diet likely to produce fatty deposits in the coronary arteries (an atherogenic diet) but remained entirely sedentary. The second group was also fed a high fat diet, but these monkeys were given controlled exercise on a treadmill during an 18 month period. The third group, which acted as a control, was fed a normal diet but also remained sedentary. At post mortem, the differences between the monkeys who exercised on the high fat diet and those who took the high fat diet *without* exercise were striking. The coronary arteries in the exercised monkeys had very much less coronary artery disease and also had arteries that were substantially wider than in the non exercising group.

One must always be cautious about extrapolating results on animals to human beings, but there does seem to be quite a lot of evidence to suggest that exercise exerts a direct benefit on the coronary arteries.

3. Blood pressure

High blood pressure (hypertension) is generally accepted to be one of the greatest risk factors for CHD. There is some evidence that exercise can help to lower blood pressure in certain individuals. Many studies show that resting blood pressures in physically fit persons are significantly lower than in those who are unfit. The response to exercise may vary considerably from one individual to another; it is likely that exercise has more effect if the blood pressure is elevated to start with. In those with normal initial blood pressure readings, exercise training appears to exert little effect. Exercise does, therefore, have a role to play in reducing high blood pressure, although the mechanisms by which this is achieved have not yet been fully elucidated. Part of the story may be the reduction in body fat which regular exercise can cause.

4. Changes in blood lipids (fat)

Blood cholesterol is a major predictor for CHD, and the higher the level the greater the risk. Conversely, HDL (high density lipoprotein) has been shown to exert a *protective* effect on the heart, possibly by removing fatty deposits from the lining of the arterial wall.

Exercise has very little effect on total cholesterol levels in the blood stream, but it does appear to have a beneficial effect on levels of HDL. A number of studies show a clear relationship between levels of HDL and activity patterns. In one study, 39 sedentary males with an average age of 33 years were studied before and after two months of jogging, walking or running three times a week. These subjects averaged a total of only five to six miles a week, nevertheless levels of HDL increased substantially from 36.9 to 55.5 per cent of total lipoprotein levels.

5. Changes in blood coagulation factors

An increased tendency for the blood to clot (coagulate) may be an important factor in causing heart attacks. There is some evidence that exercise reduces this tendency via a number of highly complex mechanisms, particularly a process known as 'fibrinolysis'. Exercise appears to increase fibrinolysis, and this in turn appears to reduce the tendency of the blood to form clots throughout the body, including the coronary arteries.

6. Reducing obesity

It is almost universally believed that regular exercise helps to reduce excess weight and maintain ideal weight, and the evidence to support this is considerable.

Exercise increases the rate at which the body burns energy (the metabolic rate) during the exercise period itself and, more importantly, for many hours afterwards. This prolonged effect of exercise on total calorie expenditure is believed by some to make exercise an important means of losing weight. However, some recent work suggests that although exercise does increase metabolic rate both during the exercise session itself and for a number of hours thereafter, the increase is actually very small. Exercise does not appear to result in much greater calorie expenditure unless it is carried out for many hours each day. It is possible that exercise promotes weight loss by mechanisms other than an increase in metabolic rate. Experience suggests that the most effective weight loss programmes use a *combination* of dietary measures and regular exercise.

Fat people are usually fat because they eat too much, take too little exercise, or both. It follows that the most appropriate way of losing weight is either to reduce the number of calories taken in or to increase the calorie expenditure. The combination of exercise and dietary measures achieves both of these aims and, although the relative contribution of exercise may be small, the combination is highly effective. Moreover, recent studies show that exercise is a very efficient means of *maintaining* ideal body weight once this has been achieved.

7. Reducing stress

Most people who take regular physical exercise will tell you that they experience a tremendous sense of wellbeing and find it an extremely effective way of controlling and sublimating the various manifestations of stress, including tension, anger and hostility. This may be important because personality characteristics such as suppressed hostility and inwardly directed anger may place the individual at a greater risk of developing heart disease. Indeed, levels of hostility may have some influence on survival. This is discussed in more detail in Chapter 8.

In the late 1950s about 1800 men between 40 and 50 years of age employed by the Western Electric Company in the USA responded to a questionnaire that included certain components designed to assess levels of hostility. Analysis of these data some 20 years later showed that of those who were rated as low on the hostility scale, about 18 per cent had died. The death rate among those on the high end of the hostility scale was 26 to 30 per cent. A similar study on a group of medical students revealed comparable results. Over a 25-year period, only about 2 to 3 per cent of the low hostility group had shown evidence of CHD, compared with 9 to 12 per cent of the high hostility group. On the basis of this evidence, anything which reduces levels of anger and hostility, including regular exercise, would appear to be beneficial.

If exercise does help to control levels of stress and promote a sense of wellbeing, the question remains as to *how* it does it. The discovery in 1974 of a group of special chemicals called 'endorphins' appeared to shed some light on this fascinating question. Endorphins have a chemical structure which is similar to morphine, and since they are produced by the body itself it was speculated that they were the body's natural painkiller or analgesic. In 1980 Pargman and Baker developed the hypothesis further and suggested that endorphins might be involved in producing some of the psychological effects of exercise. Runners, in particular, often experience a 'high', a feeling of elation which may actually be due to an altered state of consciousness. Many people who are involved in regular endurance exercise also claim that they become almost completely unaware of the environment in which they are exercising. Marathon runners sometimes say that they have no recollection of the last few miles of the race despite the enormous physiological stresses and strains they are under. It has been suggested that endorphins may help to block normal pain sensation and, at the same time, produce an altered state of consciousness. Recent studies demonstrating an increase in endorphins as result of long distance running give support to this speculation.

It is unlikely, however, that endorphins act alone and there is probably a whole family of biochemical substances responsible for these psychological effects. It is also possible that they may produce the compulsive quality of exercise which for many people can only be adequately described as an addiction. We know that people become addicted to morphine and its derivatives and so it is conceivable that certain individuals could also become addicted to their own endorphins. This might help to explain why some people who have been forced to discontinue regular exercise experience symptoms such as irritability, insomnia and depression. Are these symptoms produced by withdrawal from endorphins and other substances produced within the body as a result of exercise? Whatever the facts, the phenomenon of 'exercise dependence' is a real one and in the main would appear to be beneficial. Regular exercise is clearly an important means of sublimating stress, tension, hostility and anger and producing a greatly enhanced sense of wellbeing. Put simply, exercise makes you feel good!

The causes of stress, and the management of stress are discussed in Chapters 8 and 13 respectively.

The benefits of exercise

We can be reasonably sure that organised fitness programmes and the general promotion of exercise within the company can, over a period of time, make a significant contribution to reducing death from heart disease, especially when combined with other modifications in lifestyle, for instance stopping cigarette smoking or reducing weight. But advocates of corporate fitness programmes argue that there are many other potential benefits, including financial ones, and that physical fitness can exert a positive influence on a number of important aspects of employment, such as work performance, absenteeism, job satisfaction and recruitment. It is worthwhile considering some of these additional benefits in more detail.

Work performance

Commonsense would suggest that a fit workforce is a more efficient workforce, but to date few methodologically sound studies on fitness and job performance have been completed. A number of European studies have indicated that fit workers are more productive, but the evidence is by no means conclusive. Further studies of better quality are awaited, but there is a growing tendency to accept that physical fitness is positively related to job performance.

Absenteesism

That employee fitness should result in reduced absenteeism seems a very reasonable expectation: the better your physical condition, the less likely you are to miss work because of illness. There is some evidence to support this contention. For example, a study at Metropolitan Life Company found that those employees who participated in a fitness programme averaged 4.8 days of sickness per year as compared to 6.2 days absence per year for the members of the control group who did not exercise. Interworth Inc in North America reported a positive correlation between fitness and work attendance. Those who took part in a physical conditioning programme were absent for an average one day per year, while non-participants were absent for an average of six days per year. The Johnson & Johnson study, mentioned previously, showed that employees who participated in an exercise programme took fewer sick days than non-participants.

Reliable studies on fitness and absenteeism in the UK are not yet available, but some high quality work in other European countries does seem to support the contention that fit employees are absent less frequently than their non-exercising counterparts. There is certainly a general tendency now to take the contention seriously.

Job satisfaction

Some studies have argued that physical fitness leads to a more positive self-image and that this in turn improves job satisfaction.

It has been shown that fitness training can lead to positive changes in mood and self-perception. These changes may be mediated by endorphins and other biochemical agents acting on the brain itself. The Johnson & Johnson group carried out a survey among a group of employees and found a significant positive correlation between participation in a regular fitness programme and job satisfaction. Once again, however, there is a serious lack of controlled study in this area and no major studies have been carried out in the UK.

Stress

We discussed the use of exercise in controlling stress earlier in this chapter. There is good evidence that physical exercise can play an important part in reducing stress at work and that fitness programmes could have a positive impact on the pervasive problem of workplace stress.

Recruitment

Proponents of corporate exercise programmes and in-house facilities claim that they are a useful means of recruitment: they are a strong selling point. There is not much in the way of scientific evidence to support this contention, but it seems obvious that anyone already interested in keeping fit would favour a company which offered appropriate facilities over one that did not.

Exercise and the organisation

Corporate organisations should do everything possible to encourage increased activity patterns within the workforce and there are a number of relatively simple ways of providing such encouragement:

1. Educating about the benefits of regular, moderate exercise, in the context of general health promotion, is very important. Organising 'fitness weeks' and so on is an excellent means of doing this.

2. Groups of individuals may wish to form their own keep-fit class, running club, swimming, tennis or badminton group. These initiatives should be encouraged because, apart from providing an opportunity for regular exercise, the social dimension of such groups makes it much more likely that their members will persevere with the activity. For many people, exercising alone is extremely boring and compliance rates tend to drop off very rapidly after a few weeks.

3. Providing exercise facilities on site is a marvellous way of promoting the value of exercise within the company. The type of facilities and the range of equipment on offer will obviously vary according to the space and the financial resources available. For most companies provisions will be relatively modest – for example, one or two exercise bicycles – while for a few organisations facilities will include a swimming pool, squash and badminton courts, weight-training room and so on. We should not under-

estimate the value of even a single exercise bicycle being made available on site, because apart from being of real value in itself, it is also a symbol of the organisation's interest in promoting the health and fitness of the workforce. In fact an exercise bicycle requires very little space and happens to be an excellent way of improving cardiorespiratory efficiency.

Whatever facilities are offered, employees must be taught how to use them properly so that they can get maximum benefit from them and do not risk injuring themselves. Most reputable companies who supply exercise equipment provide some on-site training and this is best given to a named individual with responsibility to instruct the rest of the workforce.

Some organisations are unable or unwilling to provide on-site facilities but have successfully negotiated reduced subscription rates with local health clubs or gymnasiums. This is an excellent means of promoting health and exercise within the organisation and may also allow other members of the family to participate.

Summary

There is a large and compelling body of evidence to support the view that exercise has a big part to play in the prevention of disease, particularly coronary heart disease. We should therefore be encouraging the largely sedentary population in the UK to become more physically active.

Apart from its importance in primary disease prevention, there is good evidence that physical fitness leads to improvements in mood and self-perception (which affects job satisfaction), and reduction of stress. It is likely also to enhance job performance and reduce absenteeism. Finally, fitness programmes can be a recruitment attraction and bring down company healthcare costs.

CHAPTER 5

NUTRITION FOR THE NINETIES

I was born in South Wales and grew up there during the 1950s and early 1960s. Our diet then was based upon the generally acknowledged wisdom that food such as meat, eggs, fish, cheese and milk were 'good' foods, primarily because they were high in protein, whereas such foods as potatoes and bread were 'bad' because they contained large amounts of carbohydrate. The important thing about diet was that it should be high in protein and low in carbohydrate. Not much was said about fat. I virtually never saw brown bread. Rice and pasta were unheard of and no one ever drank wine. 'Meat and two veg', followed by some sort of pudding, was the order of the day, almost every day if you could afford it.

As a schoolboy, I learned that food can be classified into three major groups – protein, fat and carbohydrate – and that vitamins were also an essential part of a 'balanced diet'. In retrospect, it seems extraordinary that when I left medical school some 15 years later, having had more than six years of formal medical education, I actually knew little more about the importance of nutrition than I did as a schoolboy. This was not because I was a bad student or because I was uninterested. It was simply that, apart from discussing obvious things such as vitamin deficiencies, nutrition in its broader sense did not feature in the curriculum. Indeed, medical education left virtually the whole subject of nutrition to the dieticians, not considering it important enough to include in the training syllabus for doctors.

Of course the world has changed since then and perhaps one of the most striking developments in science and medicine during the past decade or so has been the growing recognition that what we eat can exert an important, perhaps even decisive influence, upon our health and even upon our behaviour. 'You are what you eat' is an old adage which has recently received an enormous amount of scientific support. It is hard to imagine how anyone could have avoided hearing something about the potential harmful effects of too much fat and sugar, for instance, and as a nation we are certainly much more conscious about the relationship between diet and health than we were a decade ago. This knowledge is producing changes in attitude and behaviour, although not to the same extent throughout the community, as we shall see. Unfortunately, the messages on food and nutrition which come across to the public from scientists, doctors, the food industry, the slimming industry and so on are often confusing and contradictory. Acknowledging this situation, the government set up a working party in 1973 to provide simple nutritional guidelines for the population as a whole. A similar effort was being made in other parts of the world, particularly in the United States. A committee chaired by Senator George McGovern produced a report in 1977 entitled 'Dietary Goals for the United States'. McGovern says:

> The simple fact is that our diets have changed radically within the last 50 years, with great and often very harmful effects on our health . . . too much fat, too much sugar or salt, can be and are linked directly to heart disease, cancer, obesity and stroke, among other killer diseases . . . . Those of us within government have an obligation to acknowledge this. The public wants some guidance, wants to know the truth.

In Britain, as a result of the 1973 working party, the National Advisory Committee on Nutrition Education (NACNE) was set up in 1979. The NACNE report itself was subsequently published in 1983 and another major report from the Committee on Medical Aspects of Food Policy (COMA) was published in 1984. Both of these reports represent an enormous amount of work and contain quantities of technical information. Neither of them is easily accessible to the public, but the essential message of both of them is clear: *we should eat less animal fat, processed sugar and salt, and increase our intake of fibre.*

The irony of the situation is that for most of man's time on earth the main problem with food has been the lack of it – and for many millions of people in the world that is still the case – yet now we need expert committees on food because many of us are, quite literally, eating ourselves to death. It is worthwhile looking briefly at how this extraordinary state of affairs could have come about.

Primitive man – our nutritional heritage

Ancient man evolved as a hunter-gatherer. His physical structure and his metabolism were suited to a physically active mode of existence and a mainly vegetarian diet. For virtually all of the four million years or so that man has walked this planet, he has lived mainly on plants, berries, fruit, root crops and other vegetables, with animal sources of protein such as meat and fish as a *supplement* to his diet. Hunting animals required a certain degree of social organisation, together with the necessary weapons, and could be extremely dangerous.

In scientific terms man is 'omnivorous', that is, he is able to derive energy from a very wide range of food sources, both animal and vegetable. However, most of our teeth are molars and are used for grinding rather than cutting food. The structure of our alimentary system (gut) also indicates that man is primarily a vegetarian. Therefore, although we describe man as a hunter-gatherer, the emphasis should be on the gathering rather than the hunting, the vegetable rather than the animal.

As societies grew and flourished, methods of growing crops and other vegetable sources of food were introduced, constituting a primitive form of agriculture. The vast majority of the population were landworkers, and this has remained so for virtually the whole of human history. Even as recently as 1900, approximately 60 per cent of the American population lived on farms. During this period of time the basic structure of our diet changed very little: it was a diet primarily derived from vegetable sources and, as such, contained a large amount of fibre and was low in fat, sugar and salt. What changes did occur were gradual, allowing a normal process of adaptation.

Nutrition in the 1980s

There was actually very little change in our eating habits until the period between 1700 and 1900 when the agricultural and industrial revolutions marked a profound change in man's evolution. The whole social structure was disturbed. Environmental conditions changed more radically than at any other time in the past. But the technological revolution of the twentieth century has had an even greater social and environmental impact. The industrialisation of food production has brought about a phenomenal change in our dietary habits. It would be fair to say that our diet has changed more during the past 50 to 100 years than at any other time in man's history.

Today's farming is intensified to increase food production to the maximum using ever more mechanisation and powerful chemicals such as fertilizers, insecticides, herbicides and fungicides. Food production is in the hands of enormous, often multinational, companies and profits are to a very large extent volume-related. The modern phenomenon of food processing is designed not only to enhance the appearance of the food in question, but also to substantially extend its potential shelf-life. These two aims are achieved by the use of a vast number of additives (estimated to number between 2000 and 3000) and also by the addition of other materials, in particular, fats, sugar and salt. The food manufacturer is able to offer a truly bewildering array of products which look attractive, have a long shelf-life and are easily transportable over long distances. More than three quarters of the food consumed in Britain and in the USA is processed. For example, in the USA, as recently as 1950, almost all potatoes were eaten fresh; now most (60–65 per cent) are processed, largely into frozen french fries, which, in terms of weight, have become America's most consumed vegetable. The problem is that highly processed food has lost a great deal of its original nutritional value – trace elements, minerals, vitamins, dietary fibre and protein. Not only that; the chemicals which are added to enhance appearance and preserve shelf-life, have long-term effects which are cumulative and so far unknown.

It is the speed with which our diet and eating habits have changed which is so threatening. Living organisms can adapt to changes in their environment provided that these changes occur gradually; those who are better equipped to survive, from a genetic point of view, continue to reproduce. But the huge changes in the composition and quantity of the food we eat have, in evolutionary terms, been instantaneous. Our bodies have not had sufficient time to evolve and adapt to the new diet. Those afflictions of modern society which are often called 'diseases of affluence' are more properly termed 'diseases of maladaptation'. The hunter-gather has suddenly been transformed into 'homo sedentarious' (sedentary modern man). The diet which changed very little over centuries has suddenly been replaced by what Rothman and Radford called 'the diet of industrial-capitalist man'. This diet is high in fat, high in sugar and salt, and low in fibre – exactly the converse of that of our ancestors.

We now know that diet makes such a big contribution to the twentieth-century diseases such as heart disease, cancer, obesity, diabetes, hypertension and gall stones. But what is the relationship between the various components of our diet and these diseases?

Dietary fat

Our fat consumption has increased very considerably during the past 50 to 100 years. In Britain we now eat an average of about four and a half ounces of fat per day or about 100 pounds of fat per person per year. This means that approximately 40 per cent of our total energy is currently derived from fat. Fat contains more than twice the number of calories per unit than either protein or carbohydrate, so it is a very important factor in obesity problems.

Most people are aware that consumption of meat and dairy products has increased in recent years and that they are a source of fat. People may be less aware, however, of 'hidden' fat in today's diet which is probably just as important: biscuits, cakes, chocolate, crisps, hamburgers and salami, for instance, contain very significant amounts of fat.

The chemistry of fats is immensely complex and outside the scope of this book. However, it is important to understand some basic principles about the types of fats and their relationship with blood cholesterol. Fats can be divided into two categories: saturated and polyunsaturated. Table 5.1 below indicates the differences between these two forms of fat which principally concern us here.

Table 5.1 *Characteristics of Saturated and Polyunsaturated Fats*

Saturated Fats	Polyunsaturated Fats
Solid at room temperature Primarily animal in origin Tends to increase blood cholesterol Strong association with coronary heart disease (CHD)	Liquid at room temperature Primarily vegetable in origin Tends to reduce blood cholesterol Probably protective against coronary heart disease (CHD)

Examples of saturated fats are dairy fats (cheese, butter, cream), meat fats (for example, pork, beef, mutton, lard) and processed fats (for example, biscuits, cakes, chocolate, pies, sausages, crisps). Examples of polyunsaturated fats are fish oils, sunflower oil, soya oil, safflower oil, corn (maize) oil, olive oil and some polyunsaturated margarines. It is extremely important to understand that cholesterol is only found in animal foods. Vegetables and plants contain no cholesterol at all. Generally speaking, the more animal fat a food contains, the higher the level of cholesterol. We shall see later that saturated fats are powerfully associated with the development of coronary heart disease because they tend to increase blood cholesterol. It follows that what we should be seeking is a reduction in our total saturated fat intake and an increase in our intake of polyunsaturated fats. As we shall see later, this is precisely what the NACNE and the COMA report recommend.

Fat is associated with three major forms of disease: CHD, cancer and obesity. Obesity is discussed in some detail later in this chapter, and here we focus on the relationship between fat and CHD and between fat and cancer.

Fat and heart disease

The relationship between a high intake of saturated fat and high incidence of coronary heart disease is well documented. A principal player in this process is a substance which virtually everyone will have heard of by now: cholesterol. Cholesterol is not transported freely in the blood, but is attached to another substance called low density lypoprotein (LDL). LDL and cholesterol together form a complex known as LDL-C and it is this substance which is thought to cause the damage to the lining of the arterial wall and the progressive 'furring up' of the inside of the artery known as atheroma. The basic process is thought to be as follows: a diet rich in saturated fat produces high levels of LDL-C in the bloodstream and this in turn is deposited in the lining of the arterial wall, producing the plaque or atheroma. (See Figure 5.1.) Of course this an immensely simplified account of a very complex process.

However, it does provide us with an extremely useful working model.

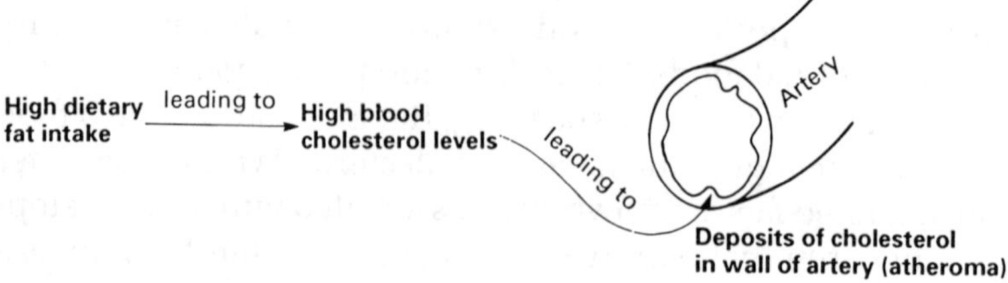

Fig. 5.1 *Origin of arterial disease due to high fat diet*

Although cholesterol is generally regarded as the villain of the piece, it is important to realise that it is in fact a vital component of the cells of the body. Chemically, cholesterol resembles an alcohol and is a member of a group of substances referred to as sterols. Cholesterol is found only in animal products. The body makes its own cholesterol, a process which can probably occur in almost every organ throughout the body except the adult brain. Most cholesterol production is, however, carried out in the liver which is capable of producing many times more cholesterol than that which is obtained through the normal diet.

There is a second form of cholesterol known as high density lipoprotein cholesterol (HDL-C). The relationship between HDL-C and diet is less clear than in the case of LDL-C. What we do know is that HDL-C exerts a powerful protective influence upon the heart and may do this by acting on the wall of the artery in such a way as to remove some of fat deposits which were laid down by the LDL-C. It follows that high levels of HDL-C in the blood should be associated with low levels of coronary heart disease, and there are many large-scale studies which prove this association. These two substances, LDL-C and HDL-C, thus appear to exert opposite effects. LDL-C causes damage to the lining of the arterial wall and the deposition of a fatty substance; HDL-C tends to remove this substance from the arterial wall and therefore exerts a measure of protection. (See Table 5.2.)

Table 5.2 *Contrasting actions of HDL-C and LDL-C*

HDL-C	LDL-C
Probably removes fatty substance from artery wall	Promotes deposition of fatty substance in artery wall
High levels of HDL-C associated with low levels of CHD	High levels of LDL-C associated with high rates of CHD

HDL-C = High Density Lipoprotein Cholesterol
LDL-C = Low Density Lipoprotein Cholesterol
CHD = Coronary Heart Disease

While the relationship between HDL-C and diet may not be straightforward, there are some factors which clearly do affect circulating levels of HDL-C. It is well known that regular vigorous exercise raises the level of HDL-C and therefore provides a measure of protection for the heart. On the other hand, smoking, overweight and obesity, inactivity, diabetes, and chronic kidney disease are all associated with low levels of HDL-C. HDL-C levels are known to be higher in women than in men and this may help to explain why there is such a difference in the incidence of heart disease between the sexes; the incidence in women is, on average, four to six times lower than in men. Whether these high levels of HDL-C are produced by female hormones is not clear. However, this mechanism is suggested by the fact that after the menopause, when levels of female hormones drop, there is a concomitant reduction in levels of HDL-C. This, in turn, is accompanied by an increasing incidence of coronary heart disease. In addition, women who take hormone replacement therapy appear to have a lower incidence of coronary heart disease than those who do not.

The cholesterol story

There can be no serious doubt that the level of cholesterol in the blood is a major determinant of coronary heart disease. Many large-scale studies have confirmed this relationship, including the Seven Countries study and the Framingham study.

The Seven Countries Study

In the early 1950s Professor Ancel Keys, a physiologist from Minneapolis, and several co-workers set up one of the most ambitious studies on risk factors for heart disease ever undertaken. Sixteen samples of men aged 40 to 59 years were drawn from seven countries: the USA, Japan, Finland, Yugoslavia, Italy, the Netherlands and Greece. Data were obtained during 1957 to 1962 on more than 12,000 men; after an initial examination all men were followed, with particular reference to the development of heart disease, and reports on the five-year and ten-year follow-up periods have been published. From this study it appears that the level of the total blood cholesterol is a key factor in determining the level of risk for the development of coronary heart disease. Furthermore, the level of cholesterol in the blood is in turn determined very largely by the amount of saturated fat in the diet. The results are summarised in Figure 5.2, which shows that levels of cholesterol are clearly associated with death rates from heart disease.

The Framingham Heart Study

The Framingham Study has gone down in history as perhaps the most important study on heart disease and the associated risk factors ever carried out. The study was established just after the Second World War because of the growing concern over an apparent upswing in death rates from heart disease in

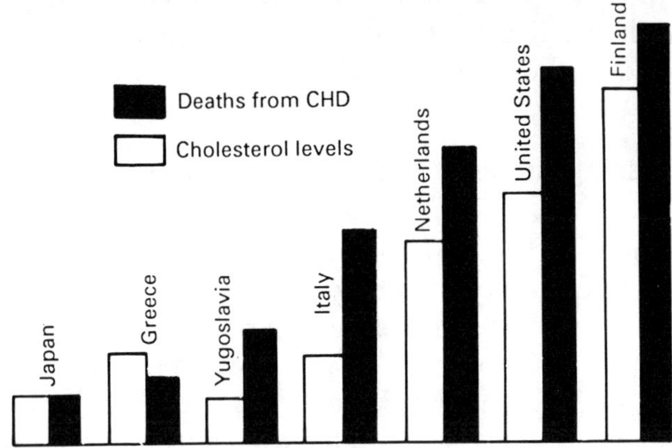

Source: Keys A (ed) 'Coronary heart disease in seven countries' Circulation *41 (Supp I 1): I–1, 1970*

Fig 5.2 *Increasing death rates from CHD with increasing levels of blood cholesterol (Seven Countries Study)*

the United States. No one really understood the cause of this epidemic, and so in 1948 a group of scientists and physicians gathered together with the aim of setting up a long-term study that would yield as much information as possible about the risk factors associated with the development of heart disease.

At that time Framingham was a small, self-contained town of about 28,000 inhabitants, situated 18 miles west of Boston. A sample of some 5000 men and women aged 30 to 59 years (including about 1500 men aged 40 to 59 years) were recruited, all of whom agreed to be examined at two-yearly intervals. Follow-up has continued to the present time, producing an enormous amount of valuable information relating the various risk factors measured at the initial and subsequent examinations to the development of coronary heart disease. As in the Seven Countries Study, the Framingham researchers have shown that those among the initial sample, who subsequently developed heart attacks or strokes, had raised initial blood cholesterol levels.

Furthermore, the Framingham Study also clearly demonstrated the *inverse* relationship between levels of HDL-C and coronary heart disease. In other words, high levels of HDL-C were associated with *low* levels of CHD.

Does lowering cholesterol make any difference?

It is established that cholesterol level is an important predictor of coronary events within a population, and that much of the cholesterol problem is related to the excessive consumption of saturated fat. However, the converse hypothesis, that lowering blood cholesterol *reduces* death rates from heart disease – which may *seem* totally reasonable – is not so easy to demonstrate. It cannot be assumed to be true.

A number of intervention trials were conducted in the 1960s and 1970s, in which attempts were made to lower blood cholesterol levels in a defined population and to follow the effect on heart disease rate over a number of years. While these studies were encouraging, they were all open to criticism and could not be used as scientific proof that lowering blood cholesterol lowers the risk of coronary heart disease.

It was against this background of frustration, confusion and disillusionment that a massive study was set up in the United States with the aim of demonstrating once and for all whether or not lowering elevated blood cholesterol would lower the risk of heart disease. The Lipid Research Clinic's Coronary Primary Prevention Trial took 12 years to complete and at more than $150 million was one of the largest and costliest medical studies ever undertaken. The results from 18 separate centres throughout the United States and Canada were coordinated at the National Heart, Lung and Blood Institute in Maryland. In all, about 480,000 American men aged 35 to 59 years were screened, yielding a high risk of 3806 men who were then followed up over a seven-year period. Eventually 193,000 clinic visits were made to produce more than one million data forms, more than 340,000 blood tests and 72,000 electrocardiograms.

In 1984, an announcement was made that was, in retrospect, a milestone in medical history: 'This study clearly shows that lowering cholesterol levels in the blood can reduce the risk of having a heart attack . . . There can no longer be any doubt that cholesterol causes heart disease.' The study estimated that a 25 per cent reduction in blood total cholesterol levels would lead to a 50 per cent reduction in the incidence of heart disease.

Other studies since have demonstrated a clear relationship between reduction in blood cholesterol levels and a reduced risk of heart disease. None of these studies is perfect but it is extremely doubtful whether, even with unlimited money, the perfect study could ever be designed. The question is, whether on the basis of the current evidence, we should be making recommendations to the population as a whole to reduce their intake of saturated fat. The answer must be an unequivocal *Yes.*

Fish oil and olive oil

These two types of oil appear to be particularly desirable components of a healthy diet.

Fish oils in fatty fish such as herrings, mackerel and sprats are rich in substances called essential fatty acids, especially eicosapentaenoic acid. These substances are thought to protect against heart attacks by reducing the tendency of the blood to form clots. This is one of the reasons why fish, which is also an excellent source of vitamins and protein, is such a healthy food.

There is quite a lot of evidence that olive oil, a mono-unsaturated oil, is better than other polyunsaturated fats for preventing heart disease. It has been known for almost 20 years that Mediterranean people, like the Greeks and Italians, have much less heart disease than we do. This observation has led to a more detailed study of the differences in dietary habits, including consumption of olive oil. Studies carried out in Spain, the United States and the Netherlands, have confirmed that olive oil does indeed confer a significant amount of protection. The oil seems to reduce the level of 'bad' (LDL) cholesterol in the blood while preserving or even increasing the level of 'good' (HDL) cholesterol.

Fat and cancer

Scientists are becoming more and more convinced that fat is a major culprit in several forms of cancer, particularly cancer of the breast, prostate and bowel.

So far as breast cancer is concerned, the incidence is high in those countries where intake of dietary fat is high. Conversely, in those countries where fat consumption is relatively low, one finds very low rates of cancer of the breast. Africans develop breast cancer only about one fifth as often as black Americans, but black and white Americans appear to have a similar level of risk. There is clearly a significant difference in fat consumption between Africa and the United States.

The relationship between fat consumption and cancer of the breast can also be demonstrated in specific communities. For example, in America breast cancer is much less common in Seventh Day Adventist women, who are mostly vegetarian and who therefore consume considerably less fat than average. A study by Dr Lawrence Kolonel on the island of Hawaii was especially interesting. The population there is composed of immigrant American and Japanese, as well as native Hawaiians. The diet there tends to reflect this mixed culture, and in 1981 Dr Kolonel was able to demonstrate that those groups of women who ate large amounts of fatty food were much more liable to develop breast cancer. The mechanism by which this may be brought about has not been defined, but it is possible that it is in some way connected with the action of certain hormones, particularly oestrogens, which are known to influence the development of breast tumours. Dr Kolonel's work in Hawaii also suggested that cancer of the prostate was more prevalent among those groups with a high fat intake.

Finally, it is very probable that excessive fat in the diet can substantially increase the risk of large bowel cancer and that adequate fibre helps to provide protection against this. The mechanism by which fat causes bowel cancer may in some way be connected with the bacteria and bile salts in the colon. Fat in the diet tends to increase the amount of bile salts found in the colon, and it is possible that these bile salts are converted into cancer-causing substances by the microbes which live in the gut. There is a lot of evidence to suggest that people with high concentrations of bile acids in their stools have the greatest risk of getting bowel cancer.

The National Research Council's report of 1982 on 'Diet, Nutrition and Cancer' presented three main sources of evidence incriminating fat as a potential cause of cancer, particularly breast, prostate and bowel cancer. First, the comparisons between different countries discussed above; second, surveys of the dietary habits of groups of patients; and third, laboratory experiments on animals which, as early as the 1940s, were already implicating fat as a cause of various cancers.

Our understanding of the relationship between fat and various forms of cancer is likely to grow considerably over the next few years, and many other forms of cancer will be probably be implicated.

Vegetables, fruit and cancer

There is growing evidence to suggest that green vegetables and citrus fruits may provide a measure of protection against cancer of various kinds. The precise mechanisms are not yet known, but it seems that the presence of vitamin A and vitamin C are important factors.

Vitamin A has attracted a great deal of attention because of the observation that it can prevent a process called 'metaplasia' in the body – an abnormal pattern of cell growth which often precedes cancer. The fact that vitamin A apparently inhibits metaplasia has led to a view among many researchers that the amount of vitamin A we eat may well help to determine whether or not we get the disease. Vitamin A is found in milk, butter, cheese, egg yolk, liver and some fatty fish. The liver oils of fish are actually the richest natural sources, but these are used as nutritional supplements rather than foods. In any event, these animal sources should only be used in moderation because they are also rich in fat which is not recommended for a healthy diet. Vitamin A can also be manufactured in the body from carotene, which is found in a wide variety of plant foods. It is chiefly found in green vegetables, such as cabbage, lettuce, broccoli and spinach, in yellow and red fruits and other vegetables.

Vitamin C is also thought to provide some protection, and, like vitamin A, it is found in fresh fruits and raw vegetables.

Fibre and disease

We should be clear about what dietary fibre actually is. It used to be called 'roughage', and it is the substance which forms the cell walls of plant foods. It is therefore the main structural component of all plants, vegetables, fruit, seeds and so on. It is not actually one substance which is common to all plants: it varies greatly from one food to another. The important thing about dietary fibre is that it is that part of our food not digested by the body.

It is estimated that between 1880 and 1960 there was a 90 per cent decline in the volume of fibre in the British diet. We now eat approximately three quarters of an ounce (20 grammes) of fibre each day. About 48 per cent of this is found in vegetables, 30 per cent in cereals and bread, 10 per cent in fruit and the remaining 12 per cent in nuts and so on.

This total consumption is extremely low compared with rural Africans whose intake is between two and four and a half ounces (50 to 120 grammes) per day. Vegetarians in the UK consume more than twice the national average of fibre – one and a half ounces (42 grammes) per day. These differences are important because, as we shall see later, there is good evidence that low-fibre diets significantly increase the risk of various diseases, including cancer.

There are two principal forms of fibre:

Insoluble fibre

This is required for a normal, smooth and regular bowel movement. It tends to promote the growth of bacteria, and makes a soft bulky stool. Insoluble fibre is found mostly in cereals but also in fruit and vegetables.

Soluble fibre

This is particularly found in oat products and beans but is also found in lentils and peas. It tends to absorb water in the stomach to form a gel and it appears to exert an important beneficial effect by lowering both blood sugar and blood cholesterol levels. This may explain why diets high in fibre are associated with lower cholesterol concentrations and therefore with a lower incidence of coronary heart disease.

Evidence of a link between lack of fibre and various diseases was initially obtained by comparing Western societies with rural communities in third world countries. These communities tend to eat a diet which is very high in fibre, whereas in the West, with our highly processed food, we tend to eat a diet rich in fat, but seriously lacking in fibre.

Lack of dietary fibre is thought to be associated with a number of diseases.

- *Large bowel cancer.* Cancer of the large bowel (colon and rectum) is the second most common cause of death from cancer in this country. There is no other cancer so strongly related to economic development and the Western way of life. The geographical distribution of large bowel cancer points strongly to changes in the pattern of diet as the cause of the disease. In societies where the intake of fibre is high, cancer of the bowel is an uncommon condition. The critical factor appears to be a combination of a low consumption of fibre and a high consumption of fats. We have seen that fat in the diet tends to increase the amount of bile salts in the bowel, upon which microbes may then act to produce carcinogens (cancer-causing substances). Fibre has the effect of reducing the activity of the microbes, and the additional bulk from fibre tends to move the waste material rapidly along the bowel. Therefore, the potentially damaging substances are in contact with the lining of the bowel for a much shorter period of time. It is clear that whereas fat in the diet may tend to cause bowel cancer, fibre in the diet tends to exert a powerful protective effect.

- *Constipation.* Two out of every five people in Britain say that they are constipated. Constipation is due to lack of fibre in the diet. About one person in five takes laxatives. About £5 million is spent on them every year. However, the best (and the most natural) way to deal with the problem is to eat more high-fibre foods. It is healthier and cheaper than taking laxatives.
- *Diverticular disease.* This is a disease of large bowel, the characteristic feature of which is small pouches in the wall of the bowel which can then become inflamed. Again, this is believed to be associated with a low-fibre diet, and increasing the fibre in the diet tends to relieve the symptoms.
- *Heart disease.* We have seen that diets high in fat tend to predispose individuals to the development of heart disease and strokes. One of the most important factors in the level of cholesterol in the blood and there is some

evidence that a high-fibre diet can help to reduce this. The most effective form of fibre is the soluble fibre found particularly in oat products and beans. A diet high in this form of fibre helps to reduce the level of cholesterol in the blood and so reduce the risk of a heart attack.

- *Diabetes.* Again, there is some evidence that diabetes is related to the Western way of life and in particular to our diet. It is very uncommon in communities with diets which are high in fibre and it may be that increasing the fibre in our diet would reduce the risk of developing it.
- *Gall stones.* The removal of the gall bladder is one of the most frequent abdominal operations in Western societies, and gall stones have become much more common during this century. They are extremely rare in third world countries and this may be connected with dietary differences.
- *Appendicitis.* This is the most frequent cause of emergency abdominal surgery in Western cultures. About one person in eight will have their appendix removed at some point in their life. Again, there is some evidence that this disease may be due to a lack of fibre in the diet since it is extremely uncommon in third world rural communities.

There are a number of other diseases, including hiatus hernia, varicose veins, irritable bowel syndrome and haemorrhoids, which may also be related to our diet, particularly a lack of fibre.

Sugar

Between 1860 and 1960 the average annual consumption of sugar approximately doubled. On average we eat nearly two pounds of sugar every week in this country, about 100 pounds of sugar per head per year. On average this means that we derive about 3400 calories per week purely from sugar. It is interesting that between 1955 and 1980, the amount of sugar bought by the packet fell by about a third. At first sight this would seem to be a most encouraging sign that we are becoming much more conscious about the need to reduce our sugar intake. However, the average consumption per head in the country has not actually changed. So where is all this extra sugar coming from? The answer, of course, is that an enormous amount of it is 'hidden' sugar. Like the hidden fat we mentioned earlier, sugar is an ingredient of a large variety of foods in which it is not obvious or expected. We know that food such as cakes and pastries have a high sugar content but we might not realise that tinned soups, tinned vegetables, pizzas, frozen prawns, sauces, pickles, tinned meats, cereals and cornflakes all contain sugar. It is in fact difficult nowadays to find any processed food which does not contain at least some sugar.

The point about sugar is that it has no nutritional value whatever. It is nothing but empty calories. No vitamins, no minerals, no fibre, no protein, nothing but energy. And this is true irrespective of whether the sugar is white or brown, whether it comes wrapped up in biscuits, cakes or fizzy drinks or is simply spooned into your tea or coffee.

Everyone knows that sugar is a direct cause of tooth decay, but does it have any other damaging effects on our health? It is often stated that sugar intake promotes heart disease, but it is in fact very doubtful that sugar makes any direct contribution to the progressive clogging up of the arteries which, as we have seen, is the cardinal feature of coronary heart disease.

By far and away the most important potential effect of eating too much sugar is obesity. Because sugar contains such a large number of unwanted and completely empty calories, it does predispose us to develop excess body fat which can lead to overweight and obesity. As we shall see later, obesity is a significant risk factor for a whole range of diseases, including high blood pressure, strokes, heart disease and diabetes. Thus it is the indirect effect of sugar, via obesity, which should alert us to the need to reduce our total sugar intake.

Salt

Salt (sodium chloride) is, of course, essential to life. All the cells that make up our body are bathed in a fluid which contains sodium and chloride in solution, in other words a salt solution. The concentration of sodium in this fluid is very carefully regulated by a series of highly complex mechanisms primarily orchestrated by the kidney.

The British eat an enormous amount of salt – on average about eight to 12 grammes per day. The important part of salt so far as our health is concerned is the sodium. Although we measure the intake of sodium chloride in grammes, we normally measure the intake of sodium itself in millimoles (mmols). One gramme of sodium chloride equals 17.1 mmols of sodium. The average daily British intake of sodium is therefore 136 to 205 mmols.

Our ancestors depended for sodium on fruits and vegetables and some occasional meat or fish which had a slightly higher sodium content. It is possible to survive on this very low sodium diet partly because of the kidneys' ability to preserve sodium and retain fluid even when in short supply. Studies carried out on primitive tribes and estimates of our ancestors' consumption agree that the natural sodium intake is no more than about 10 mmols per day. This means that we currently consume between ten and 20 times the amount required for survival. The reason why this should concern us is the fact that there is now considerable evidence that salt could be a critical factor in the development of raised blood pressure. Animal studies have shown a close relationship between salt and high blood pressure and that the relationship is especially strong when excessive consumption starts young.

It is also suggested that the current intake of salt among children is excessive and may predispose to the development of high blood pressure in later life. A recent Dutch study divided 476 newborn babies into two groups, based upon their dietary salt intake. Blood pressure was measured at a week old and then every fourth week until the babies were 25 weeks old. The results showed very clearly that the low salt group had blood pressure significantly lower than the high salt group and that the differences between these two groups increased with age.

Comparing salt intake in primitive communities, where very little sodium is eaten, with developed countries which consume much more, has shown a direct relationship between salt intake and hypertension. People, and whole societies, who do not have problems with high blood pressure have certain characteristics in common. They tend to eat large amounts of the mineral potassium and small amounts of sodium, they also tend to be slim, exercise a lot and eat small amounts of animal fat.

Because of its possible association with high blood pressure, there is an emerging view that we should all reduce our total salt intake, particularly those of us who already have raised blood pressure or who have a family history of strokes or heart attacks. We shall see later that hypertension is a major risk factor for these conditions.

While there is strong experimental and clinical evidence linking salt with high blood pressure the risk is probably not evenly distributed – it seems we are not all equally salt-sensitive. A possible mechanism is that in some individuals there is a defect in the kidneys' ability to get rid of sodium, which may predispose to the development of high blood pressure. However, this defect may be present only in about 20 per cent of the population. The problem is that there is not yet any easy way of knowing whether you are one of these more vulnerable or susceptible people. We do not currently have a simple test for salt sensitivity and the only evidence we can accumulate is circumstantial. If your mother, father, brother, or sister had elevated blood pressure then there might be a family tendency to salt-sensitivity and it would make sense to reduce your salt intake significantly. However, as the whole population eats far too much salt anyway, it makes more sense for us all to make significant reductions in our total salt intake.

This is not as easy as it sounds. The problem is that, as with fat and sugar, an enormous amount of our salt consumption is 'hidden'. Most of the salt we consume is in the manufactured and processed foods which make up so much of our diet. This means that there is not a great deal that the individual can do to reduce his salt intake unless he is prepared to make some fairly dramatic alterations to his diet as a whole. There is, however, one simple measure that we can all adopt – to use less salt in cooking and to refrain from adding salt at the table.

Overweight and obesity

'Make less thy body hence, and more thy grace;
Leave gormandising; know the grave doth gape
For thee thrice wider than for other men.'

Shakespeare, Henry IV Part II

Obesity and overweight are often identified as the biggest public health problems in the West. Certainly, if we are to judge in terms of sheer numbers, this is true. More than 50 per cent of all men and 40 per cent of all women over the age of 40 are overweight. Neither is the problem confined to middle-aged people since about 15 per cent of 16 to 19 year olds are also overweight. As a general statement we can say that, for all adult ages, approximately one third of all adults in Britain are overweight.

Obesity, which can be regarded as a more serious form of overweight (see below) affected about 6 per cent of men and 8 per cent of women over the age of 15 in 1981. This means that at that time (taking an average of 7 per cent of 40 million) 2.8 million people would have been classified as obese. According to the Royal College of Physicians, which produced these figures, being just a few pounds overweight can tip the scales against your future health.

In the United States the problem is even more prevalent. It has been variously estimated at 20 to 50 per cent of the adult population, depending on the measurement methods used and the definition of obesity chosen. Not only that, there is evidence that obesity is increasing in the United States. Data from the National Health Surveys indicate that the mean weight in men has risen from 75.3 kg in the period 1960–1962 to 78 kg in 1976–1980. In women, similar changes have occurred with the mean weight 63.5 kg (1960 to 1962) rising to 65.3 kg in the later survey. There is additional evidence that this trend is continuing and, because of the close association between obesity and a wide variety of diseases, this is clearly a matter for considerable concern.

One result of the public concern is that the yearning for a slim trim figure has become something of a national obsession and has produced a multi-million-pound slimming industry. Indeed, it is estimated that about two thirds of adult women and about one third of adult men (20 million Americans) are trying to slim at any one time. By any standards, this is big business.

As we shall see later the urge to lose weight is well founded because obesity increases the risk for a variety of diseases, including coronary heart disease, hypertension (high blood pressure), diabetes, back troubles, arthritis, chest problems and so on.

What is overweight and obesity?

'Obesity' can be defined as a condition in which there is an excessive amount of body fat. 'Overweight' is simply exceeding an 'ideal' weight. The complications come in defining the ideal.

In most countries the standards of so-called 'ideal' weight are most commonly derived from insurance data collected by the Metropolitan Life Insurance Company of New York. They have been widely used internationally simply because they represent the largest body of information relating weight to mortality. By analysing the data it is possible to relate weight to life expectancy and derive so-called 'norms' against which the remainder of the population can be measured. The problem with this is that the definition of the reference standard is arbitrary. If we were to take a different population of people who were not, as in this case, subscribers to life insurance, then it is quite possible that we would derive a completely different set of 'norms'. In other words, the reference standards do, to some extent, depend upon the population being studied. This makes it difficult to make meaningful comparisons between populations, and between studies done in different countries.

Another way of measuring overweight or obesity is by relating body weight to height. In fact, weight divided by height squared (in kg/m^2) is a convenient index of relative weight. It is convenient because it provides a single number called the Body Mass Index (or Quetelet's Index). Using this index we can define an acceptable body weight, obesity, and gross obesity:

Acceptable weight	=	Body Mass Index of 20 to 25.
Overweight	=	Body Mass Index greater than 25 but less than 30.
Obesity	=	Body Mass Index of 30 and above.
Gross obesity	=	Body Mass Index of 40 and above.

So, for example, if you are 1.86 m tall and your weight is 75 kg then your Body Mass Index is given by:

$$\frac{\text{weight}}{(\text{height})^2} = \frac{75}{3.4596} = 21.76$$

i.e. a Body Mass Index of about 22, which is well within the acceptable range.

If, on the other hand, at a height of 1.86 m your weight is 105 kg, then your Body Mass Index is given by:

$$\frac{\text{weight}}{(\text{height})^2} = \frac{105}{3.4596} = 30.35$$

In other words your Body Mass Index is above 30 and you would therefore be classified as obese.

Table 5.3 *Guidelines for body weight in adults*

Height without shoes	**Weight (kg) without clothes**			
m	**Acceptable (BMI 20–25)**	**Overweight (BMI >25<30)**	**Obese (BMI 30+)**	**(Grossly obese) (BMI 40+)**
1.45	42–53	54–63	64	(85)
1.48	42–54	55–64	65	(86)
1.50	43–55	56–65	66	(88)
1.52	44–57	58–67	68	(90)
1.54	44–58	59–69	70	(93)
1.56	45–58	59–69	70	(93)
1.58	51–64	65–76	77	(102)
1.60	52–65	66–77	78	(104)
1.62	53–66	67–78	79	(105)
1.64	54–67	68–79	80	(106)
1.66	55–69	70–82	83	(110)
1.68	56–71	72–84	85	(113)
1.70	58–73	74–87	88	(117)
1.72	59–74	75–88	89	(118)
1.74	60–75	76–89	90	(120)
1.76	62–77	78–91	92	(122)
1.78	64–79	80–94	95	(126)
1.80	65–80	81–95	96	(128)
1.82	66–82	83–97	98	(130)
1.84	67–84	85–100	101	(134)
1.86	69–86	87–102	103	(137)
1.88	71–88	89–105	106	(141)
1.90	73–90	91–107	108	(144)
1.92	75–93	94–111	112	(150)

BMI Body Mass Index
> greater than
< less than

Source: Adapted from S Truswell, ABC of Nutrition, *British Medical Journal, 1986*

Table 5.3 is based on the tables published by the Metropolitan Life Insurance Company of New York in 1959. You will see that there is no separate set of figures for males and females because modern prospective data show no significant differences between men and women when considering life expectancy and body weight. The table includes columns for Body Mass Index and you will see that acceptable weight ranges lie between 20 and 25, overweight greater than 25 but less than 30, with obesity and gross obesity starting at 30 and 40 respectively. A person classified as obese (Body Mass Index of 30 or above) is about 20 per cent above the upper end of the acceptable range of weights, and the person who is defined as being grossly obese (Body Mass Index of 40 or above) is 60 per cent above the upper end of the acceptable weight range.

So now we can define 'overweight' simply as the weight range which gives a Body Mass Index of greater than 25 but less than 30. The Index allows us to define a spectrum of weight which merges from acceptable, into overweight, into obesity and eventually into gross obesity. (See Table 5.4.) It is now a simple matter to determine whether a given weight is acceptable or not.

Table 5.4 *Classification of weight according to Body Mass Index and percentage above ideal weight*

Classification of weight	Body Mass Index (BMI)	Percentage above ideal weight
Acceptable	20–25	0
Overweight	>25<30	1–19
Obese	30+	20 or above
Grossly obese	40+	60 or above

> greater than
< less than

The consequences of overweight and obesity

There is absolutely no doubt that, over a period of years, being overweight or obese will have a significant effect on your health and will make a number of diseases more likely. These risks become serious for those who are by definition obese (Body Mass Index of 30 and above) but it is important to remember that even minor degrees of overweight can be significant. The following are a few of the conditions with which overweight and obesity are strongly associated.

- *Overweight, obesity and blood cholesterol.* Some studies have shown that there is a direct relationship between body fat, obesity and levels of blood cholesterol. In other words, the fatter you are, the higher your cholesterol level. In addition, there is also good evidence that obesity tends to lower the levels of HDL, which as we have seen earlier do provide some measure of protection for the heart. Thus by creating higher levels of cholesterol and lower levels of HDL, obesity tends to significantly increase the risk of heart disease.
- *Overweight, obesity and high blood pressure (hypertension).* Even small degrees of overweight can cause a significant rise in blood pressure and, as we have seen previously, hypertension is a major risk factor for heart disease and stroke. Conversely, reducing weight tends to lower blood pressure. Data from the Chicago Coronary Prevention Evaluation Programme, by Stamler and his colleagues, showed a clear correlation between weight reduction and blood pressure reduction over a five year period.
- *Overweight, obesity and heart disease.* Being overweight increases your risk of dying from heart disease in several ways. As we have seen, obesity tends to be associated with high blood cholesterol levels, high blood pressure and also lack of exercise. For these reasons people who are obese tend to have significantly higher degrees of risk than those who are not. The question of whether obesity and overweight are *independent* risk factors for heart disease (apart, that is, from their association with other known risk factors) is

more difficult to evaluate. In general terms, for men aged 35 to 44 a 10 per cent increase in body weight increases the chance of heart attack by approximately 38 per cent. A 20 per cent increase in weight (that is obesity) increases the risk of a heart attack by 86 per cent.

- *Overweight, obesity and cancer.* From a statistical point of view, as we have seen earlier, there is a greater risk of developing certain types of cancer in those who are overweight. In men, cancer of the large bowel and the prostate may well be associated with obesity and, in women, obesity is also associated with cancer of the breast. Again, it is difficult to determine whether it is the obesity as such which promotes the increased risk, or whether it is simply that the diets of the obese tend to be high in fat and, therefore, high in calories. From a preventive point of view the distinction is an academic one.
- *Overweight, obesity and lung disease.* One of the most common symptoms of obesity is breathlessness after very mild exertion, such as carrying shopping bags or walking upstairs. There is no doubt that the capacity of the lungs to function normally deteriorates as obesity increases. In addition, if you are overweight you are a much greater anaesthetic risk if you have to have surgery. You are also more likely to develop a post-operative chest infection.
- *Overweight, obesity and arthritis.* The chronic and progressive degeneration of the joints (osteoarthritis) which affects us all to some degree sooner or later, is accelerated in the obese. The likely explanation of this is purely mechanical – the body simply has to transport a much greater weight than it was designed for and the consequent wear and tear on the joints is that much greater. Symptoms associated with arthritis often improve when the individual loses weight.
- *Overweight, obesity and diabetes.* So-called 'maturity onset' diabetes is a common condition in middle and

old age. It is very often associated with obesity, and reducing weight can be a useful (and in some cases the only necessary) form of treatment.

- *Overweight, obesity and gall bladder disease.* As we have seen, high fat diets tend to predispose to diseases of the gall bladder, particularly gall stones. People who consume high fat diets, also tend to be overweight or obese which may explain the association between gall bladder disease and obesity. Again, whether this is an independent factor or merely the consequence of diet is difficult to determine.

Apart from the conditions listed above there are many other consequences of being overweight or obese. General fatigue is an extremely common complaint among overweight people and while it most often does not indicate anything serious, it does substantially reduce the quality of life for many individuals. Back problems including slipped discs are much more liable to be a problem in the overweight and obese.

Finally, we should not forget the potential psychological consequences of being overweight or obese. For many fat people the need to lose weight becomes almost an obsession and their continued failure to do so leads to feelings of guilt, inadequacy and sometimes real depressive illness. This depression and loss of self-esteem may be made even more acute in a society which is increasingly obsessed with image and personal appearance and which seems always to attribute success, power and sex appeal to the slim rather than to the fat. The social pressure to lose weight emphasises the need to improve personal appearance, rather than to improve *health*.

Summary

It will by now be obvious that what we eat is of enormous importance in determining our health and future wellbeing.

With the average Western person eating about half a ton of food a year, it is a subject which should concern us all, not only for our own sakes but also for our children. Some of the recent studies on the prevalence of coronary heart disease risk factors and eating habits in children have been nothing short of alarming. Obesity, inactivity, high levels of blood cholesterol, cigarette smoking and so on are quite common in children, and, if this pattern of behaviour continues into adult life, then we can regard these children as 'incubators' of disease for the future. It is essential that we should encourage healthy eating habits at an early stage and continue them throughout life. This is never easy, particularly with a food industry which recognises the importance of the children's market and targets them specifically. It is very difficult to maintain influence on our children's eating habits when they are at school or at play, but we can ensure that they are set the example of healthy eating at home. We can summarise our discussion thus far by saying that many of our modern diseases are clearly associated with poor dietary habits, specifically diets which are high in fat, sugar, salt and low in fibre. Current evidence would suggest that reducing the amount of fat, sugar and salt in our diets and at the same time substantially increasing our consumption of fibre could do much to prevent these diseases occuring in the first place. At this point it would be useful to be more specific about the degree of change required; the recommendations of the NACNE and the COMA reports on fat, fibre, sugar and salt in our diet can be summarised as follows:

- *Fat.* In general terms fat consumption should be cut by one quarter, to 30 per cent of total energy intake from the current level of about 40 per cent.

  Saturated fat consumption should be cut by nearly one half to 10 per cent of the total energy intake from the present level of 18 per cent.

  As far as polyunsaturated fat is concerned there was no specific recommendation to increase the amount eaten. However, the ratio of polyunsaturated to saturated fat (the P/S ratio) will rise because of a cut in saturated fat.

Fish oil and olive oil may both afford a significant measure of protection against heart disease, and therefore their inclusion in the normal diet is to be encouraged.

- *Fibre.* The consumption of fibre should be increased by approximately one half to 30 grammes a day from the present level of 20 grammes a day.

  Green vegetables and fresh fruit are an essential component of healthy eating and their inclusion may help to provide a measure of protection against various forms of cancer.
- *Sugar.* Sugar consumption should be cut by about one half so that it represents 10 per cent of the total energy intake from the current level of 20 per cent.
- *Salt.* We should reduce our salt consumption by approximately half to 5 grammes (80 mmols of sodium) per day.

These principles are extremely important because they form the framework within which a healthy eating policy not only for an individual but also for a whole workforce can be constructed.

About 1 billion canteen meals are consumed each year, and company catering can be a hazard for employees though it is seen as a benefit. Company canteens and restaurants have been slower to offer healthier menus in Britain than the USA, but there is a growing trend in the right direction, largely as a result of evidence linking poor diet with various diseases.

It is important not to try and dictate what employees should eat because this is often counter productive. Completely removing chips, sausages and pies from the menu is *not* the way to encourage healthier eating. The key to success is to have other healthier foods such as baked potatoes, rice and salads on offer at the same time, so that people are able to make healthier choices if they wish. Canteen staff themselves can do a great deal to encourage healthy eating and their cooperation is extremely important.

Current eating trends

Many people in this country are changing their eating habits for the better. Manufacturers too are beginning to recognise the new market opportunities being created by the changing awareness of healthy eating.

But attitudes and behaviour are not changing at the same pace across all social groups. The Health and Lifestyle Survey of 1987 reported on the physical and mental health attitudes of a random sample of more than 9000 British adults. The section on dietary habits and eating patterns is very revealing. For example, despite the very considerable publicity about the importance of dietary fibre for a healthy diet, knowledge about the fibre content of specific foods was quite poor, and this raises the question of whether people really understand the term 'dietary fibre'. The survey also showed very clearly that even among people shown to be aware of the importance of dietary fibre, less than half the men and less than two thirds of the women actually consumed brown bread. This demonstrates a point that we shall be discussing later – that there is not necessarily any correlation between knowledge and behaviour. On the question of fatty foods, there was considerable socio-economic variation. The consumption of chips and similar foods was considerably – three times – more common in manual and unskilled groups than in professional and managerial occupations.

CHAPTER 6

SMOKING

'. . .this is a plague and a mischief, a violent purger of goods, lands, health, hellish, devilish and damned tobacco, the ruin and overthrow of body and soul.'

Robert Burton (1577–1640) *Anatomy of Melancholy*

The introduction of tobacco to Great Britain is traditionally attributed to Sir Walter Raleigh, but in fact it was already being cultivated in England as early as 1573 following experiments in Portugal by the French ambassador to Lisbon, Jean Nicot. It was Nicot who apparently sent the first tobacco seeds to the Queen of France, Catherine De Medici, in the mid-sixteenth century. Nicot was said to have believed that tobacco smoke had medicinal properties, and he may have derived this view from the North American Indians who probably introduced it to Portuguese traders. Many North American Indian tribes had used tobacco for centuries, believing in its healing properties, and they used it extensively in their tribal rituals. Members of the royal courts in Europe began to experiment with tobacco, although pipe smoking and snuff taking were the most popular forms of tobacco use in the eighteenth century. Its advocates extolled its virtues in the enhancement of social and sexual life. It is said to have been the Turks, during the Crimean War in the 1850s, who taught the British how to handle cigarettes, and the patenting of Bonsac's cigarette machine in 1880 facilitated the mass production of cigarettes, the basis of the modern tobacco industry. It was, of course, from Jean Nicot that the word 'nicotine' was derived.

Tobacco smoke

Tobacco smoke is produced by incomplete combustion of the

tobacco leaf and composed of gases and vapours in which drops are dispersed. Chemical analysis of the smoke reveals more than 3800 different chemical compounds, more than 50 of which are known to be carcinogenic (cancer-causing agents). Mainstream smoke is the smoke inhaled directly by the smoker and then exhaled into the environment. Sidestream smoke, on the other hand, is the smoke emitted from the smouldering tip of the cigarette in between puffs. Harmful substances such as carbon monoxide, ammonia, tar, nitrosamines and polycyclic monocarbons tend to be more concentrated in sidestream smoke than mainstream smoke. The smoker, of course, inhales both mainstream and sidestream smoke while the non-smoker (passive smoker) inhales the sidestream and the exhaled mainstream smoke of the smoker. When tobacco smoke is inhaled into the lungs, the effect of cooling causes it to condense and this, in turn, produces the dark brown tar which collects in the lungs of smokers.

Apart from the cancer-causing agents, cigarette smoke also contains nicotine. Nicotine has a damaging effect on the heart and inhalation of cigarette smoke causes an increase in blood pressure, an increase in heart rate and certain undesirable changes in the composition of blood. Even in very small doses it is a stimulant and in large doses it can be fatal. Nicotine plays an important but not exclusive role in maintaining the smoking habit.

Tobacco smoke also contains carbon monoxide. This is a gas which combines with haemoglobin, the substance which normally carries oxygen in the blood. The problem is that carbon monoxide has a 200-times greater propensity to combine with haemoglobin than oxygen and so it replaces the normal oxygen-carrying capacity of the blood to a significant extent. The combination of carbon monoxide and haemoglobin is known as carboxyhaemoglobin (COHB). In non-smokers COHB levels are only 1 per cent, but in smokers they may be as high as 10 to 15 per cent. This is important because there is some evidence that it is the COHB in the blood which causes damage to the lining of the arterial wall and helps to produce the build up of fatty substance in the arteries known as atheroma.

Cigar and pipe smoking

The risk of developing lung cancer as a result of smoking pipes or cigars is significantly less than in cigarette smokers. However, pipe and cigar smokers have a high incidence of cancer of the mouth, lip, larynx and gullet, and this is probably due to the different way the smoke is inhaled. Pipe and cigar smokers tend to inhale less smoke into the lungs and hold more of the smoke in the mouth and upper respiratory tract. Of male cigarette smokers in the United Kingdom, 78 per cent state that they inhale a lot or a fair amount, compared with 38 per cent of cigar smokers and 16 per cent of pipe smokers.

So far as coronary heart disease is concerned, most studies show that there is only a very slightly increased risk for cigar and pipe smokers when compared with non-smokers. It is certainly nothing like the degree of risk incurred by cigarette smoking.

As a general rule, it is better to be a pipe or cigar smoker than a cigarette smoker, although it is obviously better not to smoke at all. It is often said that cigarette smokers would be better switching to pipe or cigar smoking. This is probably true, but the problem is that when cigarette smokers switch to pipes or cigars, they still tend to inhale the smoke. In other words, they tend to smoke cigars or pipes in exactly the same way as they smoked cigarettes, and this is probably just as harmful.

Trends in cigarette smoking

It is extremely encouraging that during the past decade or so, there has been a substantial reduction in the numbers of adults smoking cigarettes in this country. This is due to a number of factors, the most important of which is probably mounting social pressure upon smokers to give up. Smoking is increasingly being regarded as an anti-social habit, particularly as the evidence accumulates connecting passive smoking with disease. The other obvious factor is the increasing awareness of the dangers to health of cigarette smoking. Both factors have been considerably strengthened by the efforts of organisations such as Action on Smoking and Health and the Health Education Authority.

Historically, cigarette consumption in males and females has differed considerably. Men began to smoke cigarettes in the early part of this century and their consumption rose steadily to reach a peak of about 12 cigarettes per adult male per day in the early 1940s. After the war, there was a slight decrease in consumption, but thereafter it remained fairly constant until the early 1970s when a dramatic decrease began. (See Figure 6.1.)

Smoking in women became more common in the war years and rose during the 1950s and 1960s to reach a peak of about seven cigarettes per day in the early 1970s. Since then, tobacco consumption in women has fallen, but less dramatically than in men.

The General Household Survey of 1986 confirmed the decline of cigarette smoking in both men and women. (See Figure 6.2.)

Although the trend is in the right direction, one person in every three continues to smoke, and about 40 per cent of these may be classified as 'heavy' smokers with a daily consumption of 20 cigarettes or more. The General Household Survey also revealed worrying disparities between social groups. Smoking habits, like eating habits, have changed more among professional and managerial groups than among the semi-skilled or the unskilled. (See Figure 6.3.) Not surprisingly, there are more ex-smokers among the professional and managerial groups than among the manual workers.

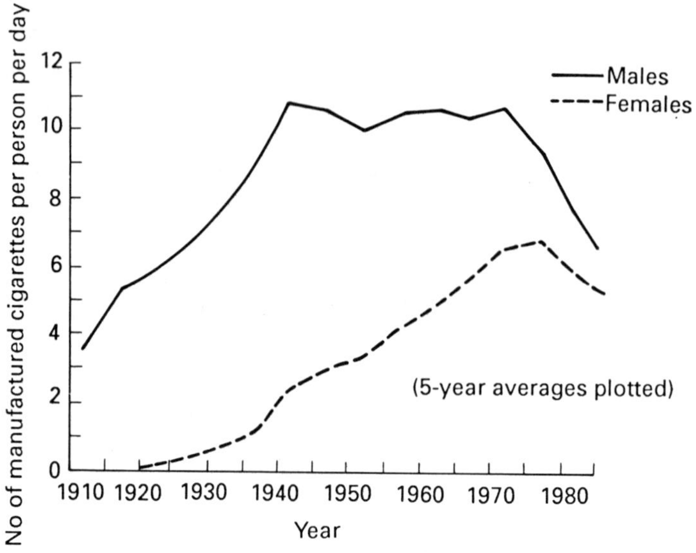

With permission of the Cancer Research Campaign

Fig. 6.1 *Daily consumption of cigarettes in the UK by men and women aged 15 and over: 1910–1985*

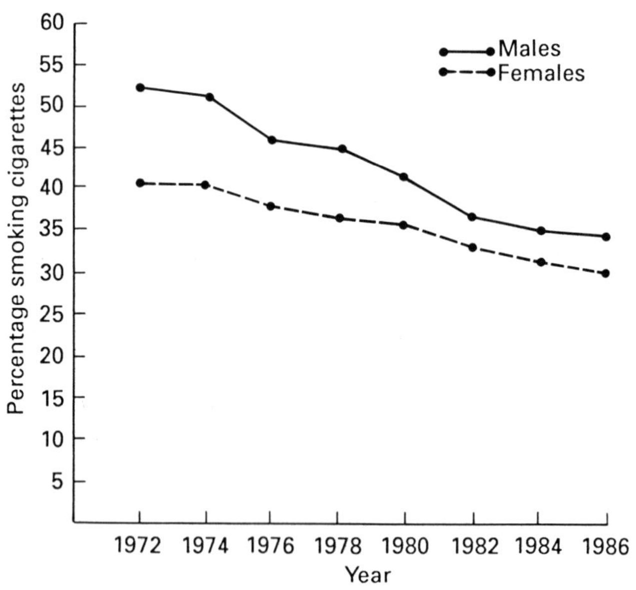

Fig. 6.2 *Cigarette smoking by sex: 1972–1986*

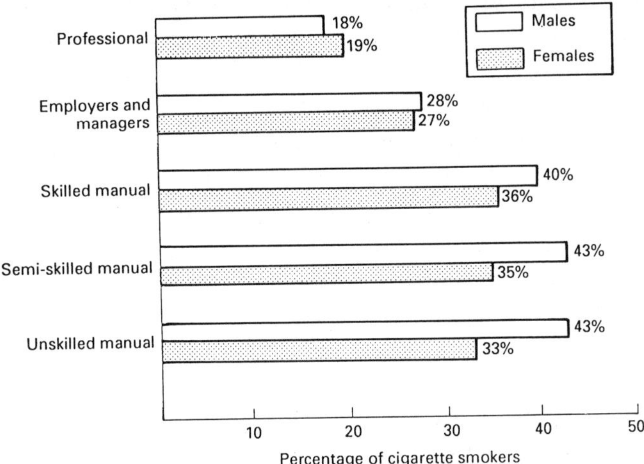

Fig. 6.3 *Prevalence of cigarette smoking by sex and socio-economic group (1986)*

An interesting pattern emerges when we look at the actual numbers of cigarettes smoked by each group. While the professional group tends on average to smoke fewer cigarettes each week than smokers in the other groups, smokers in the employers and managerial groups average more than any other group. When we look at female smokers, exactly the same is true. (See Table 6.1.) In other words, although there are fewer smokers in the employers' and managerial groups than in the manual groups, those who do continue to smoke have a heavier cigarette consumption than any other group.

Table 6.1 *Average weekly cigarette consumption 1986*

Socio-economic group	Average weekly cigarette consumption	
	Male	**Female**
Professional	85	82
Employers and managers	130	101
Skilled manual	118	99
Semi-skilled manual	114	101
Unskilled manual	110	92

While adults are smoking less, unfortunately children are smoking more. A survey in 1984 which examined cigarette smoking among school-age children showed prevalence to increase with age. By their last compulsory year at school, 31 per cent of boys and 28 per cent of girls were regular smokers.

Average cigarette consumption among these children was 49 cigarettes per week, although one in four pupils smoked 70 cigarettes per week. From these figures it was calculated that pupils in England and Wales spent between £1.2 and 1.6 million each week on between 20 and 25 million cigarettes. Moreover, the percentage of regular smokers in 1984 was two points higher than that recorded in 1982, suggesting that the upward trend in continuing.

So far as the type of cigarettes being smoked is concerned, there is a significant difference between the sexes. Women are far more likely than men to smoke lower tar brands (23 per cent of women compared with 14 per cent of men). Women are also less likely to smoke middle tar brands (36 per cent compared with 45 per cent of men). About similar proportions of men and women smoke low to middle tar brands of manufactured cigarettes. This is important because there is some evidence that reducing the tar content of the cigarette makes it less likely to cause lung cancer. However, it is unlikely that smoking a low tar cigarette will reduce the risk of developing heart disease and other forms of lung disease.

Smoking-associated disease

There are three groups of diseases most often associated with cigarette smoking. They are cancer, cardiovascular disease (disease of the heart and arteries) and lung diseases, such as bronchitis and emphysema.

Cancer

Nowadays most people are aware of the association between smoking and lung cancer but smoking is also strongly related to cancer of the throat, mouth, oesophagus (gullet), pancreas and bladder. Indeed, Doll and Peto have estimated that 30 per cent of all cancers are attributable to tobacco smoking.

Britain has the highest lung cancer rate in the world. The disease increased among men a staggering 50-fold between 1911 and 1970, an epidemic which closely followed the increasing popularity of cigarette smoking this century. In 1986, about 40,000 people died from lung cancer, 28,500 males and 11,500 females. Women are probably as susceptible to smoking-related diseases as men, but because, as a group, they have not been smoking for so long, the amount of disease related to cigarette smoking in women is below that in men, although it is rising. Indeed women in Britain, America and many other countries are now suffering a significant increase in lung cancer which has paralleled the trend for cigarette smoking in women. In the 1930s convincing experimental evidence relating smoking to cancer in laboratory animals was produced, and in the 1950s and 1960s Doll and Hill and others published epidemiological evidence strongly linking cigarette smoking with lung cancer and other forms of lung disease. They showed very clearly that lung cancer in non-smokers is a rare disease.

In the United States there are about 120,000 lung cancer deaths each year, 85,000 in males and 35,000 in females. As in Britain, about one heavy smoker in every five can expect to get lung cancer.

The risk of developing lung cancer is related to the number of cigarettes smoked. Individuals who smoke one packet of cigarettes a day are ten times more at risk than non-smokers; those who smoke two packets of cigarettes a day are 25 times more at risk than non-smokers. Workers in the asbestos and uranium mining industries who smoke cigarettes run an especially high risk of lung cancer. It is estimated that about 90 per cent of lung cancer deaths are attributable to cigarette smoking and are preventable.

Cigarette, cigar and pipe smoking in men is associated with

about a fivefold increased risk of developing cancer of the mouth, the oesophagus and the throat.

Recovery rates from any of these cancers are very low. Only about 7 or 8 per cent of lung cancer patients are alive five years after diagnosis; 7 per cent of patients with cancer of the oesophagus; and 4 per cent of patients with pancreatic cancer.

Passive smoking

Smoking harms not just smokers themselves but those who live, work and socialise with them. Concern has recently been growing about the effects of 'passive smoking' – inhaling the smoke from other people's cigarettes or pipes. In 1988 the United Kingdom Independent Scientific Committee on Smoking and Health, chaired by Sir Peter Froggatt, stated that 'passive smoking is associated with an increased risk of lung cancer in non-smokers'. It recommended that further publicity should be given to this fact.

The extent of the risk is probably in the region of 10 to 30 per cent. This means that if the risk of lung cancer in non-smokers is about 10 per 100,000 per year, the risk in an exposed group is 11 to 13 per 100,000 per year. It means that several hundred of the 40,000 or so deaths from lung cancer each year may well be caused by passive smoking.

A number of studies have looked at lung cancer deaths among non-smoking wives of smoking husbands. In Japan 73 per cent of men but only 15 per cent of women smoke, but the death rates from lung cancer are similar in both sexes. Research carried out by Takeshi Hirayama of the National Cancer Centre Research Institute in Tokyo provided some alarming statistics. During a 14-year period he followed the medical histories of more than 90,000 non-smoking women married to smokers. He concluded that women married to pipe smokers are about 1.6 times likely to die from lung cancer than those married to non-smokers, and women whose husbands smoke 20 or more cigarettes per day are more than twice as likely to die from the disease. This links up with the knowledge that the risk of lung cancer increases with the number of cigarattes smoked and also with the period of exposure.

A study carried out in Greece supports Hirayama's conclusions, although it puts the level of risk slightly higher: 2.4 times the risk for wives of smokers of less than one packet of cigarettes a day and 3.4 times the risk for wives of smokers of more than one packet a day.

Of course tobacco smoke can cause a whole variety of much less serious ailments. It can cause eye, nose and throat irritation and headaches, and many people simply find the smell unpleasant. Increased levels of nicotine and carbon monoxide can be found in the bloodstream of non-smokers as a result of exposure to cigarette smoke. Adults who already have some form of cardiac or respiratory disease, such as angina or asthma, find that exposure to cigarette smoke can cause a significant worsening of symptoms.

There is compelling evidence that cigarette smoking in pregnancy can adversely affect the developing child. Babies of mothers who smoke weigh, on average, 200 grammes less than babies born to non-smoking mothers. Mothers who smoke greatly increase the chance of their pregnancies ending in spontaneous abortion, stillbirth or neonatal death. Maternal smoking in pregnancy has also been identified as a risk factor in cot death. On top of all this, children of parents who smoke get more chest infections in the early years of life.

Cardiovascular disease

Two forms of cardiovascular disease are strongly associated with cigarette smoking: coronary heart disease, which we discussed earlier, and peripheral arterial disease. Smoking has been estimated to account for about a quarter of all coronary deaths in men and women under the age of 65.

The overall risk for smokers of death from coronary heart disease is probably at least twice as that for non-smokers. The risk increases with the number of cigarettes smoked.

Women who smoke cigarettes and use oral contraceptives increase their risk of suffering a heart attack more than five times.

The mechanism by which cigarette smoke helps to promote the development of coronary heart disease is not yet clear. It is almost certainly due to a combination of factors including direct damage to the lining of the arterial wall, increasing the 'stickiness' of the blood itself, lowering high density lipoprotein (HDL) which, as we have seen, is an important protective factor, and increasing blood pressure via the direct effects of nicotine.

In peripheral arterial disease the same degenerative process which occurs in CHD affects other major arteries of the body, particularly the abdomen and the limbs. Progressive build-up of fatty plaque (atheroma) reduces the blood supply to the legs. This can produce serious problems which sometimes require major surgical intervention. There is no question that cigarette smoking is an important factor in causing this damage to the peripheral arteries.

Chronic bronchitis, emphysema and other lung conditions

Cigarette smoking is the most important factor in the development of lung diseases other than cancer, particularly chronic obstructive airways disease (bronchitis) and emphysema. Emphysema is a condition in which the air sacs within the lungs become abnormally dilated and rigid and prone to infection. The vast majority of bronchitis and emphysema sufferers are smokers. As with cancer, the risk of these diseases increases with the number of cigarettes smoked and the length of exposure. In 1986 about 33,000 people died of bronchitis in Britain and about 90 per cent of this disease could be attributable directly to cigarette smoking.

Economic considerations

More than 100,000 people a year die prematurely as a direct result of smoking (more than all the civilian casualties in Britain in World War Two). On average, smokers are seven times more likely to die between the ages of 20 and 40 than are non-smokers. Overall death rates for smokers are between 70 per cent and 270 per cent higher than non-smokers.

Most of the studies relating to the direct costs of smoking have been done in the United States. In 1980, the American Health Foundation estimated that medical costs and loss of productivity due to smoking amounted to $47 billion – $36 billion in lost productivity and $11 billion in medical costs.

Out of a total of 350 million working days lost in the UK each year through sickness, the Royal College of Physicians has estimated that about 50 million are lost as a direct result of cigarette smoking. Not only is there a much greater likelihood of disease among smokers; if they do not have to enter hospital, smokers are much more likely to develop complications in the post-operative period and may therefore take longer to recover and to return to work.

It is difficult to estimate the cost of replacing pesonnel who die or become disabled prematurely. It can cost an employer £250,000 or more to replace, say, a 45-year-old executive; it depends on how much the company has invested in the executive's training and on the position held. In 1985 the Rank Xerox Organisation estimated that the cost to the company from the loss of one executive would be about £415,000.

Absenteeism rates show that employees who smoke average about 50 per cent more sick leave than their non-smoking counterparts. The National Centre for Health Statistics in the United States estimates that smokers have a 50 per cent greater chance of being hospitalised than their non-smoking colleagues. They also estimate that, in contrast to non-smokers, smokers spend nearly 150 million more days in bed and 81,000 more days off the job. A study published in

Northern Ireland by Action on Smoking and Health (ASH) showed that smokers averaged 7.3 excess days absence per year, from smoking induced illness. This level of absenteeism has a significant impact upon the financial performance of large and small companies, especially when a replacement has to be paid to carry out the work of the absentee.

There is some evidence to suggest that smokers are less productive on a day-to-day basis than their non-smoking colleagues. Some attempts have been made to estimate the amount of time taken up by smoking itself. Results vary from about eight minutes per day to 15 to 30 per hour. Of course, it depends entirely on the job, but in some cases work simply stops when a cigarette is lit. Some studies in the United States have shown that smokers lose anything between 2 and 10 per cent of their efficiency. In themselves these may not seem terribly significant, but in the aggregate they can produce a measurable difference between the performance of smokers and non-smokers. Smokers may also be less productive if the smoke causes eye irritation and reduced concentration.

Increased insurance costs

Company health insurance premiums reflect the fact that smoking makes a significant contribution to premature death and disease. Some insurance companies in the United States offer lower rates to companies with non-smoking policies or to small firms where none of the employees smokes. The American Council of Life Insurance estimates that employers spend almost $300 extra per smoker each year in health insurance claims alone. Other forms of insurance – against accidents, for instance – may also be affected; one American Council of Life Insurance study found the total accident rate among smokers to be twice that of non-smokers – accidents due mainly to loss of attention, eye irritation and coughing.

Damage to property

Damage from fire is probably the most serious risk of smoking in the workplace. Organisations which implemented non-smoking policies should be in a strong position to negotiate significant reductions in their insurance premiums.

Apart from the risk of fire damage, cigarette smoking can damage rugs, floors, equipment, furniture, curtains and so on. Another American Council of Life Insurance study estimates that a non-smoking policy in the workplace can save employers at least $500 per smoker per year in replacement of furnishings and equipment. One firm which implemented a non-smoking policy reduced their monthly maintenance contract fee by £350. The Merle Norman Cosmetic Company made workstations smoke-free, and they saved about £10,000 in the first year in reduced house-keeping costs alone.

Ventilation

Smoking also places a tremendous load on the company's air-conditioning system. Some studies in the United States indicate that the air-conditioning level required to clear a smoke-filled room may be up to six times that needed in a room where smoking does not occur. This has a significant implication for costs of maintenance and replacement of such equipment.

In addition to all this, we should not ignore the potential impact of cigarette smoking on non-smokers. Non-smokers also may suffer ill-health, loss of productivity, increased absenteeism and poor work performance due to the cigarette smoke which they may inhale from other people's cigarettes.

In an important US study of 10,000 non-smoking office workers, about 50 per cent said they had difficulty working near smokers, and a further 36 per cent said they regularly had to move from their workstations to escape the smoke.

About one in seven of those who had difficulty working near smokers said they had been absent from work during the previous 12 months because of the effects of tobacco smoke around them at work. It is absolutely clear from these studies that for non-smokers too, tobacco smoke in the workplace can have a significant negative impact on performance, productivity and attendance at work.

It is clear for all these reasons that smoking in the workplace can have a major impact upon the productivity of the individuals involved and therefore upon the financial performance of the company. A study carried out in Washington, estimated that each smoker cost his or her employer more than $4000 a year, as follows:

- Absenteeism runs at 2.2 more days each year, at a cost of about $110 per day.
- Medical care benefits are used 50 per cent more frequently by smokers than non-smokers, at an annual cost of at least $130.
- Earnings are lost to the employer because of the smoker's sickness and/or early death, at a cost of $765.
- Accidents cost an estimated $45, fire insurance costs go up an estimated $45.
- Loss of productivity through smoking breaks and so on is estimated at $1,820.
- Damage or maintenance for smoking pollution costs $1000.

Some workers regard this figure as being much too high and that the real costs are actually much less. A New York study estimated that the annual cost of smoking to the average employer in January 1980 ranged from $336 to $601 per average smoking employee.

In any event, whether the true cost is $600 or $4000, the point is that smoking makes a significant adverse medical and financial impact which employers cannot afford to ignore.

In 1986 ASH published a study which had been carried out

in 1984 in Northern Ireland to assess the economic consequences of smoking. The total estimated costs to employers from loss of productivity due to smoking-related illness and death, time spent on the smoking ritual itself and the cost of fire damage due to smoking, were over £100 million. When one thinks that Northern Ireland has less than 3 per cent of the United Kingdom population, the cost of smoking to the employer is simply enormous.

On the other side of the economic coin are the government's tax revenue from tobacco and the vested interests of the tobacco companies.

Even 25 years after the Royal College of Physicians' report on the dangers of smoking and the vast amount of evidence relating smoking to premature death and disability, the tobacco industry is still reluctant to accept any proven link between cigarette smoking and disease. The industry has been particularly interested in denying the dangers of passive smoking. This is a key issue. In a free society if people choose to adopt certain forms of behaviour which can damage their health or even kill them prematurely, that is their right. But the accumulating evidence that inhaling cigarette smoke can do significant damage to non-smokers, including the unborn child of smoking mothers, has seriously challenged the 'freedom to smoke' argument and given new impetus for change. Nevertheless, advertising and promotion of tobacco continues on a huge scale, pushing the idea of smoking as a desirable and prestigious activity linked with performance, status, sexual prowess and sophistication. The tobacco industry probably spends over £100 million per year – more than £2 million per week on advertising. (The Health Education Authority's budget for the whole of 1985 was less than £10 million and this had to be used on trying to prevent a whole range of diseases not just those which are tobacco associated.)

There is a powerful argument that preventing cigarette smoking would ruin the economy. It is backed up by some work done by the DHSS and the central statistical office in the 1970s, suggesting that a significant reduction in cigarette smoking would eventually cost the country more than could be gained through such an effort. There are two main reasons for this. First, the government currently collects about £5.5 billion per year in tobacco taxation so smoking is an important source of revenue. Second, if more people survive into old age as a result of not smoking, the bill for retirement pensions goes up. The loss of revenue and the cost of more pensions together outweigh, in economic terms, the cost of caring for the victims of smoking. Moreover, the tobacco industry itself is a big employer and no government wants to create unemployment. Other industries, for example the advertising and newspaper industry, also depend to a considerable extent upon the tobacco industry for revenue.

These factors may partly explain why the government has made only limited efforts to control the promotion of tobacco: we have the government health warning on each packet of cigarettes and we have also seen cigarette advertising banned on television. The problem is that these sorts of measures are having more impact among the middle class and better educated groups than among the lower social class groups in whom smoking-related diseases are actually more prevalent. It is clear that we must now focus our efforts much more upon the manual and unskilled groups of people in this country; we need to ensure that the health education message is clearly understandable and accessible to them.

Attitudes towards workplace smoking restrictions

In the United States the National Survey of Worksite Health Promotion Activities carried out in 1985 indicated that about 36 per cent of worksites had smoking control activities of one form or another. Of these companies, more than three-quarters had a formal non-smoking policy. Large companies in the USA which offer smoking cessation programmes include IBM, Rank Xerox, DuPont, Johnson & Johnson, the Campbell Soup Company, AT & T and Boeing.

It is extremely encouraging that this has also become a growing trend in the UK; companies include Boots, Marks & Spencer, ICI, Texaco, Iceland Frozen Foods, Mars Confectionary and Ethicon. In 1984, the London Underground was declared a non-smoking area. The Glasgow and Newcastle underground systems are also non-smoking areas. British Rail offers between two thirds and three quarters non-smoking accommodation on most of its trains.

A growing number of cinemas, pubs, hotels and restaurants also ban smoking. Thistle Hotels and Crest Hotels reserve a proportion of their bedrooms solely for non-smokers and Thorn EMI recently conducted a trial ban in 26 of its 106 ABC cinemas and may later extend this.

The most important thing which is reflected in all these trends is the significant change in public attitude towards cigarette smoking as a health hazard – among smokers as well as non-smokers. A 1986 MORI poll conducted on behalf of ASH and the Cancer Research Campaign was very revealing:

- Among smokers interviewed, less than 25 per cent were in favour of smoking being allowed in all areas of the workplace.
- Of all workers, both non-smokers and smokers, only one worker in seven believed that smoking should be allowed in all areas of the workplace.
- 80 per cent of workers (smokers and non-smokers) would agree to some form of smoking restrictions at work.
- 85 per cent of smokers and 77 per cent of non-smokers believed that smoking should either be allowed only in specific designated areas, or not permitted at all.
- Two-thirds of all those interviewed believed that smoking should be restricted to smoking areas.
- About three-quarters of the sample believed that companies should have specific policies on smoking.

The important point about this survey and others like it, is that the idea or the excuse that smokers themselves will not support a smoking policy and will therefore create difficulties in the workplace, is simply not supported by the evidence. Indeed, the converse is the case, and employers should be reassured by these data and be much more enthusiastic about introducing smoking policies into the working environment. That smokers on the whole support such policies is not really surprising: most of them would like to give the habit up and are therefore likely to agree to any changes in the working environment which might help them to do so.

Introducing a smoking policy in the workplace

The Health Education Authority, in conjunction with other organisations such as ASH, has produced an excellent guide to good practice on workplace smoking. The document not only covers the basic design and implementation of the policy, including legal issues, but also provides some interesting case studies. In view of this there is little point in covering the subject in any detail here, but these are the main steps involved in introducing a worksite smoking policy.

Step 1. Preparing for action

The first thing to do is to set up a working group responsible for the initial design and drafting of the policy. Members of this group should represent a broad constituency of opinion within the organisation, and should include managers and trade union representatives, smokers and non-smokers.

Every employee should be notified of the company's aims and objectives in establishing a workplace smoking policy, particularly in relation to the many health issues involved. Then the working group should ascertain the views of all employees at every level within the company, and the most effective way of doing this is to do a survey using a brief questionnaire. The questionnaire results must be collated and analysed.

Step 2. Drafting the policy

In the light of the information obtained in Step 1, the working party should draft an outline policy. 'Off-the-peg' policies cannot be used since programmes must be carefully tailored to fit the needs of the particular organisation and reflect the broad consensus of opinion in its workforce.

The main body of the policy will be given over to defining those areas in the workplace which will be designated as 'no-smoking' areas. However, while the underlying principle should be to make non-smoking the norm, it is extremely important not to be too draconian, or to seem to victimise smokers. A decision simply to ban smoking everywhere could be counterproductive and might incur the hostility of both smokers and non-smokers within the workforce. From an industrial relations perspective, the best approach is to protect non-smokers, but also preserve the right of smokers to continue to smoke. The important thing is to give employees themselves the opportunity to decide where smoking should or should not be allowed. Most current policies forbid smoking in the following areas:

- where it constitutes a health and safety risk
- meeting/conference rooms
- toilets, corridors and lifts
- reception areas
- interview rooms
- staff dining room/restaurant and coffee lounge (except where there are designated smoking areas).

Instead of applying one policy for the whole company, some organisations have allowed each office, department or section to make its own decisions, usually on the basis of an 'opt in', or an 'opt out' scheme. In the former, smoking is allowed until one or more persons object; in the latter, unless everyone is in agreement that it should be allowed, smoking is banned in all working areas.

Step 3. Implementing the policy

When the policy has been drafted and agreed, all staff must be made aware of its existence and its content. A sufficient period should be allowed for any further comments to be voiced, and the policy should ideally be phased in over a two- to three-month period. This allows smokers to adjust to the new rules and get used to using the designated smoking areas only. 'No-smoking' signs need to be put up in appropriate places. Consideration should be given to setting up a 'quit-smoking' programme simultaneously. This will provide valuable additional incentive for those who genuinely wish to give up.

These three steps provide the basic framework for introducing a smoking policy in the workplace. At all stages it is extremely important not to condemn smokers, but rather to emphasise the positive benefits of the policy, particularly from a health and environmental perspective. Potential benefits for the company as a whole include:

- A reduction in the risk of premature death and disability among the workforce
- Reduced absenteeism and increased productivity
- Lower maintenance costs and increased safety performance
- Fewer accidents
- Improved morale especially among non-smokers.

Summary

We have seen that smoking is the leading cause of cancer in

Britain and many other countries and is strongly associated with a variety of other crippling and killing diseases. In recent years there has been a significant decline in the number of individuals in this country who continue to smoke, although this trend has not been observed equally across all social groups. In contrast, there has been a worrying increase in the number of cigarette smokers among children. The cost of cigarette smoking to industry is enormous but government revenue from taxation is also huge. Many companies and organisations now realise that the introduction of a smoking policy in the workplace makes medical and financial sense.

CHAPTER 7

ALCOHOL AND DRUG ABUSE

On 24 March 1989, the oil tanker Exxon Valdez struck an iceberg in Prince William Sound, Alaska, and more than ten million gallons of crude oil were released into the surrounding waters. This event marked the beginning of one of the worst ecological disasters in American history, the consequences of which are likely to remain with us for decades to come. If, as has been suggested, the master of the ship was under the influence of alcohol at the time, there can be no more striking example of how alcohol abuse can have consequences not only for the individual, but for society as a whole.

Chapter 4 presented some basic facts about the scale of alcohol-related problems in this country – facts which may not be as dramatic or newsworthy as the Exxon Valdez incident, but which demonstrate beyond doubt that alcohol abuse is one of the most serious health issues of our time.

In this chapter we discuss the use and abuse of alcohol in more detail, particularly with respect to some of its physiological effects and its physical and psychological consequences.

What is alcohol?

Chemically, ethyl alcohol is a colourless liquid with the formula C_2H_5OH. From a medical point of view it is a drug which has a depressant effect upon the brain and reflexes; as with any other drug, it has certain side effects and is potentially addictive. In fact alcohol can create both a physical and a psychological dependence and has a well-defined (and sometimes dramatic) withdrawal syndrome. In large enough doses, it can exert a direct toxic effect on virtually every organ in the body.

Why do we drink alcohol?

The effects of alcohol have been known for thousands of years and its consumption (and presumably abuse) has been a major feature in the cultures of our ancestors and part of our social and religious traditions. Man has always sought escape from his predicament and alcohol has been used so much because the raw material required for its production is cheap and readily available.

Given that 95 per cent of the adult population in Great Britain use alcohol regularly and are prepared to pay more than £35 million each day for it, there must be some fairly compelling reasons for drinking it. So deeply ingrained has the use of this drug in our culture become, that we tend to notice it most when it is *not* available. A wedding reception without alcohol would be regarded as a very strange event indeed by most of us. We use alcohol to ease conversation on social occasions, to celebrate promotion, to relax after a dreadful day at the office, to toast success and achievement, to celebrate a birth, and sometimes to console ourselves in moments of disappointment, depression or grief.

There seem to be three main reasons which go at least some way to explaining why people drink. Most people drink for a combination of these reasons. The first is that many people actually enjoy the characteristics and the taste of the drink itself. In the case of the wine expert, the appreciation of bouquet, texture, balance and taste becomes almost an art form. At a less exalted level, many of us enjoy wine or other alcoholic drinks as a complement to food.

Second, we tend to use alcohol in situations where there may be a certain amount of pressure to do so. Social groups of all kinds impose a certain pressure to conform, and this often means drinking. Then there is pressure from the advertisers who encourage us to believe that alcohol is in some way associated with sophistication, wit, sex appeal and power. The non-drinker is, by implication, boring, humourless, narrow-minded and sexless.

The third reason why we drink is to experience the effects of alcohol as a drug. Alcohol has a direct depressant action on the brain and nervous system, and, depending upon the dosage, can affect our conscious level in such a way as to produce a wide spectrum of behaviour and sensations, ranging from pleasant relaxation and sociability, through garrulousness and aggression to drunken oblivion and even death. Many people use alcohol to induce feelings and behaviour patterns at one end of this spectrum, and a few people drink just to get blind drunk in the shortest time possible.

But, as a society, we are paying an increasingly heavy price for the pleasures and benefits of drinking. Alcohol-related problems are on the increase and some estimates have suggested that as many as one in ten of the total adult population are either drinking to excess or have a definite problem related to alcohol. Moreover, a recent study published in the *British Medical Journal* suggested that there may be as many as 28,000 deaths each year in England and Wales associated with alcohol consumption.

Factors affecting general consumption

Cost

Without any question, the single most important factor influencing alcohol consumption is cost. Cost and consumption are inversely related: the cheaper alcohol becomes, the more we drink. Between 1978 and 1982 when, for the first time in 30 years, the real cost of alcohol increased, there was a 20 per cent reduction in hospital admissions for alcohol-related disease, a 4 per cent reduction in deaths from cirrhosis of the liver and a 16 per cent reduction in convictions for drunkenness.

There is also no doubt that the single most powerful weapon in the fight against alcohol abuse is alcohol taxation. This is why there is so much fear about what will happen to our alcohol consumption in 1992 when the abolition of European tax differentials is expected to reduce the price of a bottle of spirits by £2.37, wine by 70p and a pint of beer by 15p. With this price reduction alcohol consumption could go up about 28 per cent and alcohol-related problems will follow suit. In the light of the current evidence, any government which fails to increase alcohol taxation at a level at least in line with inflation, is allowing political expediency to take precedence over public health.

Availability

Another major factor in the rise in alcohol consumption is its increasing availability. Licensing laws were introduced into this country by Lloyd George's government at the beginning of the First World War to curb excessive drinking by munition workers. The result was a reduction in beer consumption of almost two thirds and a reduction in the consumption of spirits by more than a half between 1914 and 1918. Since that time we have seen a steady liberalisation in licensing laws and a dramatic increase in availability of alcohol through newsagents, general stores and supermarkets. The consumption of alcohol has increased, and with it all forms of alcohol-related problems.

Advertising

While cost and availability are clearly related to levels of consumption, the effectiveness of alcohol advertising is difficult to prove. The alcohol industry has shown considerable ingenuity and creativity in its advertising and sales campaigns and an ability to target specific sections of the population, particularly the young. The industry probably spends in excess of £200 million per year on advertising.

Who develops alcohol problems?

It is a common misconception that there are two types of drinkers: those who drink 'socially' or 'normally', and those who are alcoholics. The popular image of the alcoholic is someone who is dirty, unkempt, drunk most of the time and who sleeps under bridges in cardboard boxes, or on park benches. There are indeed many individuals who do live this way and who at the same time have serious alcohol dependency problems. But the vast majority of individuals with drink problems are in regular gainful employment – a fact which has enormous implications for industry. Some occupations appear to have a much higher incidence of alcohol-related problems than others. As might be expected, publicans have by far the highest rate of alcohol liver disease and therefore the highest death rate from cirrhosis. (See Table 7.1.)

Alcohol dependency is a condition which can afflict anyone, irrespective of income, social class, occupation, sex or age. Recent evidence suggests that anyone who drinks heavily enough for a sufficient period of time can become alcohol dependent. But there is also evidence that some of us may be more at risk than others. Individuals who use alcohol primarily for its drug effects, for example, in order to cope with social pressures, financial worries or family problems may be particularly vulnerable to addiction. We also know that an individual whose friends drink heavily or whose leisure activities are mainly focused around drinking, is also at increased risk. There may be a genetic factor involved too; we know that the children of alcoholics, even when they are not brought up by their biological parents are more likely

Table 7.1 *Occupations with the highest mortality from cirrhosis of the liver*

Occupation title	Rank in top 20
Publicans, innkeepers	1
Deck, engineering officers and pilots, ship	2
Barmen, barmaids	3
Deck and engine room ratings, barge and boatmen	4
Fishermen	5
Proprietors and managers boarding houses and hotels	6
Finance, insurance brokers, financial agents	7
Restaurateurs	8
Lorry drivers' mates, van guards	9
Cooks	10
Shunters, pointsmen	11
Winders, reelers	12
Electrical engineers (so described)	13
Authors, journalists and related workers	14
Medical practitioners (qualified)	15
Garage proprietors	16
Signalmen and crossing keepers, railways	17
Maids, valets and related service workers	18
Tobacco preparers and products makers	19
Metallurgists	20

With permission of OPCS

to develop alcohol addiction than the children of non-alcoholics. It has been debated for many years whether alcohol addiction is a disease or whether it is just a manifestation of an inadequate personality structure. It is still not certain whether the alcoholic personality actually exists, but what is clear is that alcoholics do develop particular personality characteristics as a *result* of alcohol dependence. These characteristics include emotional instability, a low tolerance to stress, feelings of insecurity, isolation and depression.

The message here is very clear. The line between reasonable alcohol consumption and an excessive consumption is a lot thinner than many people think, and the slippery slope between social drinking and true alcohol addiction is a lot steeper.

What is an alcoholic?

The term 'alcoholic' has never been adequately defined and is potentially misleading. It is part of the mistaken view that there are only two classes of drinkers – normal drinkers (like you and me) and alcoholics (someone else). Moreover, 'alcoholics' are thought of as individuals with a condition which is entirely self-inflicted, and it therefore carries a strong social stigma. For these reasons, the term 'dependent drinker' is preferred.

To understand the nature of alcohol addiction, it is essential to realise that there is no clear distinction between the social drinker and the heavy drinker, between the heavy drinker and the problem drinker, or between the problem drinker and the dependent drinker. These terms are merely convenient points of description along a spectrum of alcohol consumption and its related problems. (See Figure 7.1.)

So how much alcohol is safe?

One of the difficulties in suggesting safe levels of alcohol consumption for both men and women is that individuals vary enormously. There is now a general consensus on safe levels of consumption, but it must be borne in mind that the more an individual drinks, the more likely she or he is to suffer from alcohol-related problems.

It is useful to quantify alcohol in terms of *standard units*, thus:

One standard unit = One half pint of beer
One measure of spirit
One glass of wine
One measure of vermouth, sherry, and so on.

So if you drink two pints of beer and one gin and tonic, you will have consumed a total of five units. If you did this seven

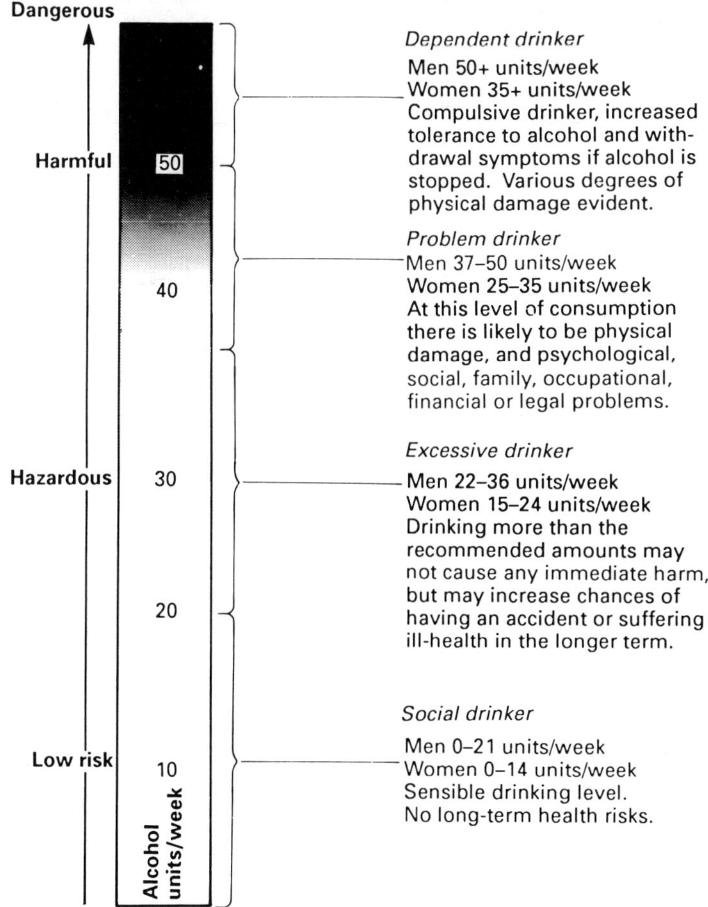

Fig. 7.1 *Alcohol problems related to consumption*

nights a week, your weekly consumption would be 35 units. Using the guidelines indicated in Table 7.2 you can estimate daily and weekly consumption.

Sensible limits of drinking are as follows:

Men: Not more than **21** units a week including two or three days without any alcohol.

Table 7.2 *Alcohol content of various drinks in units. Alcohol strength shown in ( ) is the percentage of alcohol by volume (the figure given on most wine labels) and should not be confused with 'proof' levels.*

Drink	Amount	Units
Beer and lager		
Ordinary strength bitter or draught lager (3–4%)	1 pint	2
	½ pint	1
Special draught bitter (5%)	1 pint	3
	½ pint	1½
Strong lager (e.g. Carlsberg Special: 7%)	1 can	4
Cider		
Ordinary strength cider (4–5%)	1 pint	3
	½ pint	1½
Strong cider (e.g. Strongbow: 5%)	1 can	3
Spirits e.g. whisky, rum, gin, vodka etc.		
1 pub measure of spirits (38%)	1/6 gill	1
1 small home measure of spirits (38%)	¼ gill	1½
1 bottle of spirits (38%)	1 bottle	30
Wine and champagne		
Low strength table wine (8%)	1 glass	1
	1 bottle	6
Ordinary strength table wine (12%)	1 glass	1½
	1 bottle	9
Strong table wine (16%)	1 glass	2
Champagne or other sparkling wine (12%)	1 glass	1½
	1 bottle	9
Sherry and fortified wine e.g. port, vermouth etc.		
Sherry (16%)	1 glass	1
	1 bottle	13
Vermouth (e.g. Martini, Cinzano: 15%)	1 glass	1
	1 bottle	11
Port (20%)	1 glass	1
	1 bottle	15

Women: Not more than **14** units a week including two or three days without any alcohol.

Drinking at this level does not appear to be associated with any long-term health hazards.

Is alcohol beneficial in any way?

There has been considerable debate in recent years as to whether small amounts of alcohol may actually be beneficial. The controversy centres around a V-shaped graph which many studies plotting deaths from heart disease against alcohol consumption have shown to exist. In other words, those who do not drink at all appear to be at a slightly greater risk of dying from heart disease than those who drink moderately. (See Figure 7.2.) A possible explanation is suggested by the association, shown in many studies, between moderate alcohol consumption and higher levels of HDL cholesterol which, as we have seen, appears to help prevent the build up of fatty deposits in the coronary arteries.

Critics of these studies claim that the non-drinking category is swelled by many former drinkers who have given up for health reasons. These individuals might already have sustained damage from their drinking and would therefore

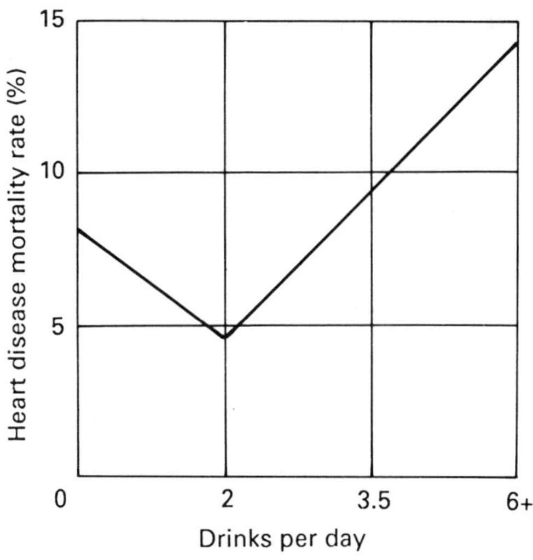

Fig. 7.2 *Alcohol consumption (drinks per day) and deaths from heart disease*

show higher mortality rates than lifelong teetotallers. If this were taken into account, it is argued, the graph would show a straight line indicating a direct relationship between alcohol consumption and mortality. Teetotallers and 'safe' drinkers would show the lowest mortality.

The argument will probably remain unresolved for some time to come. In the meantime, it may be a measure of comfort to those of us who do enjoy an occasional drink.

What happens when we drink?

When we drink, alcohol passes virtually unchanged from the mouth into the stomach; about 20 per cent is immediately absorbed into the bloodstream through the stomach wall; the remaining 80 per cent passes into the small gut where it is absorbed into the bloodstream. The alcohol then passes to the liver where a special enzyme called alcohol dehydrogenase removes the alcohol from the blood and breaks it down into waste products. Alcohol dehydrogenase appears to have evolved specifically to fulfil this purpose – a powerful testimony to man's long relationship with alcohol. After the liver has removed some of the alcohol, the remainder passes around the rest of the body where it can affect virtually every organ with which it comes into contact.

The exact amount of alcohol in the bloodstream at any point depends on a number of factors:

- The amount of alcohol which has been drunk: obviously the more you drink the higher the blood levels of alcohol.
- The weight of the drinker: the same amount of alcohol has a greater effect on a light person than on a heavy person. This is obviously because the concentration of alcohol in the bloodstream is greater in a smaller person.
- The sex of the drinker: a given amount of alcohol has a greater effect on a woman than on a man. This is partly

because on average women are lighter than men, but also because their chemistry is different. Roughly 60 per cent of body weight in men is water, but in women it is only 50 per cent. Again, therefore, any alcohol present in the body is more concentrated.

- Eating food: The presence of food in the stomach tends to slow down the rate of absorption of alcohol into the bloodstream, and this is why drinking on an empty stomach is not a good thing.
- Speed of drinking: obviously, the faster the alcohol is taken, the more rapidly blood level of alcohol rise.

Blood alcohol concentration (BAC) is measured in milligrams of alcohol per 100 millilitres of blood (mg/100ml). One unit of alcohol increases the blood alcohol level within the first hour by about 15mg per 100ml in a man, and by nearly 20mg per 100ml in a woman. In a healthy person alcohol is removed from the body by the liver at a rate of about 15mg per 100ml an hour.

This means that if someone goes to bed at midnight, with a blood alcohol level of 200mg per 100ml (having drunk, say, seven pints of beer) and then drives to work at seven o'clock the next morning, he is likely to have a blood alcohol level well over the legal limit – 80mg per 100ml. This is because in the seven-hour period his body will have removed about 105mg per 100ml of alcohol, leaving him with as much as 95mg per 100ml still in his bloodstream. For the first few hours of his working day he is significantly more likely to be involved in an accident and his overall judgement will be poorer than normal. Indeed, it will be well after noon before the residual alcohol is finally removed from his bloodstream.

As the level of alcohol in the blood rises, it begins to exert a depressant effect upon the brain, producing disturbances in thought, physical coordination, memory, judgement and sensation. Even relatively small amounts of alcohol can exert a negative impact on these brain functions which is why, after only a few drinks, people begin to slur their speech and stagger, and driving becomes dangerous.

Some may be surprised that alcohol is a depressant; the common perception is that it is a stimulant. The explanation for this can be found in the way in which alcohol affects brain function. The human brain has three discernible levels of function. (See Table 7.3.) From an evolutionary point of view, the highest centres are much more recent than the intermediate or lower centres which are the 'older' and more primitive parts of the brain. Alcohol tends to affect the highest centres first, thus allowing the functions of the lower and intermediate centres, responsible for more primitive impulses and emotions, to express themselves. This produces the boisterousness, aggression, lack of inhibition and over-confidence which gives the impression that alcohol is having a stimulating effect.

A summary of the effects of different levels of blood alcohol concentration on behaviour is given in Table 7.4.

Table 7.3 *Alcohol and brain function*

Alcohol depresses the higher centres first thus allowing the expression of intermediate centre patterns of behaviour	
Higher centres	Advanced planning Complex reasoning Calculation Sophisticated skills Memory Perception Impulse control
Intermediate centres	Emotion Aggression Hunger Thirst Sex
Lower centres	Consciousness Breathing Heart rate

Table 7.4 *Blood alcohol concentration: effects and consequences*

BAC (mg/100ml)	Effects and consequences
30 mg %	No obvious outward effects, but increased likelihood of an accident.
50 mg %	Relaxed and happy, but less vigilant, poorer judgement and more accident prone.
80 mg %	Confident, talkative, more emotional and less inhibited. Slower reaction times and twice as likely to have a road accident. Loss of driving licence if caught.
120 mg %	Much less inhibited, over-confident, more impulsive as intermediate centres express themselves. Five times more likely to have a road accident.
150 mg %	Loss of self-control, slurred speech, impaired thinking and reasoning more evident. Aggressive and over-emotional. Ten times more likely to have a road accident.
200 mg %	Slurred speech, staggering, memory loss, belligerent or depressed and self-pitying. Over twenty times more likely to have a road accident.
300 mg %	Extremely drunk, difficult to remain upright, speech very difficult. Conciousness now threatened.
400 mg %	Oblivion, sleep, lower centres of brain seriously threatened. Dangerous situation.
500 mg %	Lower centres controlling breathing and heart rate now affected. Death possible.
600 mg %	DEATH PROBABLE

What damage does alcohol do the body?

There is scarcely an organ or system in the body which is not affected by prolonged alcohol intake. (See Figure 7.3.) The body has a remarkable capacity to recover from the acute effects of alcohol, but with high intake over a sufficiently long period, damage is progressive and irreversible.

Alcohol has a direct toxic effect on the liver which is not

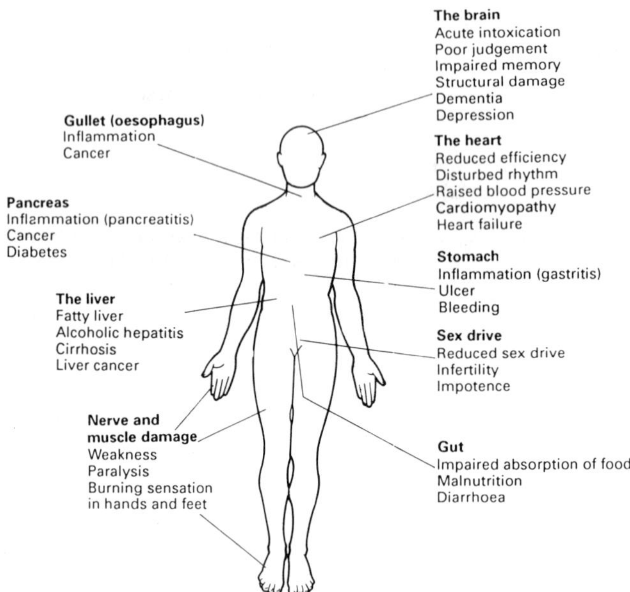

Fig. 7.3 *Physical effects of prolonged alcohol abuse*

surprising since the liver is one of the body's main defence mechanisms against alcohol. The liver's job is to convert what we eat and drink into usable nutrition while getting rid of unwanted substances. The earliest and the most common sign of liver distress as a result of alcohol intake, is 'fatty liver'. This condition is reversible if the patient stops drinking alcohol; if not, inflammation of the liver (hepatitis) may ensue. About half of those who suffer from alcoholic hepatitis go on to develop cirrhosis which is scarring of the liver, and this may eventually lead to death. Those who have been seriously alcohol-dependent for more than ten years have about a 10 per cent chance of dying from cirrhosis. In advanced cases, an individual may get drunk on relatively small amounts of alcohol because the damage to the liver means that most of the alcohol passes through without being removed.

A serious complication of cirrhosis is liver cell cancer, and this develops in anything between 3 and 11 per cent of people with alcoholic cirrhosis.

Alcohol has a direct effect on the heart, reducing the strength of the heartbeat and sometimes causing disturbances in the rhythm of the heart. Heavy and continued alcohol consumption may cause specific disease of the heart muscle (cardiomyopathy) leading to heart failure. High alcohol intake is also an important and common cause of raised blood pressure which usually reverts to normal on cessation of alcohol.

Alcohol abuse is also often associated with malnutrition. This is because the alcohol-dependent individual often has reduced total food intake, impaired digestion and absorption of food, and inadequate intake of essential vitamins. Thiamine, nicotinic acid, pyridoxine, and vitamin B12 are all essential for normal brain function; deficiencies can lead to serious nerve and brain damage, including dementia.

Alcohol and cancer

Any form of alcohol taken in large amounts increases the risk of cancer, particularly in smokers. Heavy drinkers have a significantly greater risk of developing cancer of the mouth, the throat and the gullet. Liver cancer, as a major complication of cirrhosis was mentioned above, and recent evidence suggests that heavy drinkers may be as much as three times more likely to develop cancer of the pancreas.

Some recent evidence suggests, for reasons that are not entirely clear, that heavy alcohol consumption in women may lead to a much higher risk of developing breast cancer. Researchers at the National Cancer Institute in the United States reported a 50 to 100 per cent increase in risk, depending upon the level of consumption. The evidence is at least sufficient to recommend that women who are already at risk of developing breast cancer (with a family history, for instance) should drink much less.

Alcohol and pregnancy

Excessive consumption of alcohol in pregnancy can permanently damage the unborn child, leading to abnormal facial features, low birth weight, abnormal growth, mental retardation and a variety of other abnormalities including heart, kidney and skeletal defects. This combination of abnormalities is sometimes referred to as 'foetal alcohol syndrome'.

Women who drink heavily have about a twofold increase in the risk of spontaneous abortion. There is some evidence that even very small amounts of alcohol during pregnancy may have a detrimental effect upon the developing child. For this reason, and until further studies clarify the situation, it is probably sensible for women who are pregnant or plan to become pregnant to abstain from alcohol or at least to restrict their consumption to an occasional drink, especially in the first three months of pregnancy.

Summary

We have now looked at the social consequences of alcohol abuse (Chapter 3) and, in this chapter, we have seen some of its physical and psychological effects. We shall later (Chapter 11) discuss the worksite implications of alcohol abuse. Figure 7.4 summarises the acute and chronic effects of alcohol in each of these areas. The bi-directional arrows between the acute and chronic effects indicates that they considerably overlap.

Drug abuse

Although alcohol is a drug, the difference between it and other forms of drug abuse is sufficient to warrant them being

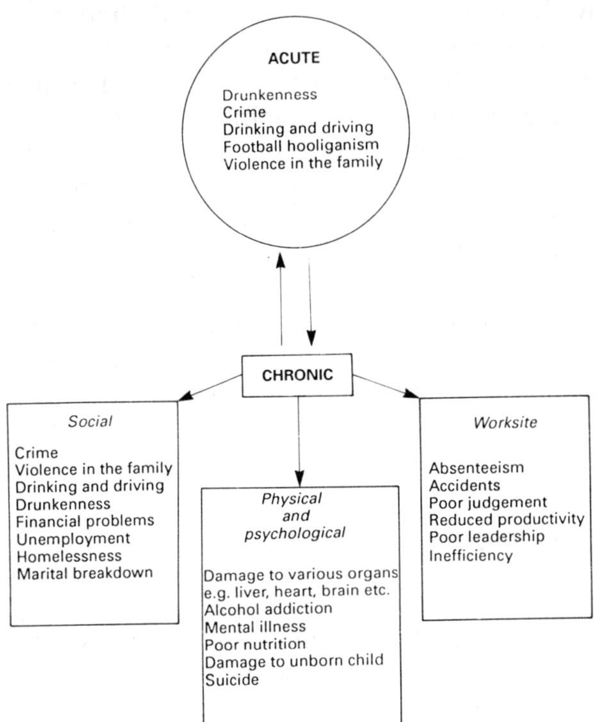

Fig. 7.4 *Alcohol-related problems*

treated separately. The main difference, of course, is that alcohol is legal and is used extensively throughout all areas of our society. Society does not disapprove of the use of alcohol, only of its abuse. Other forms of drug abuse tend to involve the use of illegal substances and although this may be accepted by some social subgroups, it remains a cultural taboo for most of our society. Corporate organisations tend to have a very different attitude towards the alcohol abuser compared to the drug abuser, reflecting a much greater level of disapproval of the latter by society as a whole. Disapproval is reinforced by a perception that drug users are part of some form of counter-culture and that the use of drugs is symbolic of a rejection of social values, a rebellion against authority and a wish to overturn the status quo. Although the consequences of alcohol abuse can be far more devastating, it is still, to some extent, the devil we know. The drug abuse problem may seem more menacing because it in some way threatens the very fabric and substance of our society.

This sense of menace and foreboding is compounded by the advent of AIDS, since intravenous transmission of the virus via infected needles is now one of the main sources of infection.

Although the use of illegal drugs such as heroin and cocaine (and all their derivatives) is on the increase, dependency on legally prescribed drugs such as tranquillisers or the various types of solvents is much more widespread.

Drugs and the workplace

As with alcohol, the main consequences of drug abuse are absenteeism, loss of efficiency, erratic behaviour, loss of productivity, mistakes and accidents. Even mild tranquillisers have an impact upon judgement and reaction time sufficient to place the individual at a much greater risk of serious injury or even death.

Drug abuse and the law

The most important legislation pertaining to the use of drugs, is the Misuse of Drugs Act, 1971. The Act divides drugs into a number of groups according to their perceived danger. Classes A and B are the ones most relevant to our discussion:

Class A includes cocaine, the opiate derivatives including morphine and diamorphine (heroin); and various hallucinogens such as LSD.

Class B includes cannabis, amphetamines and barbituates.

These are the principal illegal drugs which any employer is likely to encounter, and it is an offence to possess, to supply or to produce them. Moreover, it is an offence knowingly to permit them to be used, kept or supplied on the premises. This means that, from a legal point of view, no employer can ignore the existence of illegal drugs in the workplace.

Some drugs, of course, are obtained entirely legally, and the law in relation to these is slightly different. So far as the use of solvents such as glues and cleaning fluids is concerned, possession is not illegal; being intoxicated by them is. So far as the group of drugs known as the benzodiazepines, which includes most of the common tranquillisers such as Valium and Librium, is concerned, it is an offence to offer them to a person for whom they have not been prescribed.

The legal obligations of employers in relation to drug abuse by employees have not been adequately clarified. Some industrial tribunals have held that dismissal on the grounds of excessive alcohol consumption is not reasonable in situations where the employer has made no previous efforts to warn or to help the employee. Whether this will prove to be the case in relation to other forms of drug abuse is difficult to predict. Very few cases have so far been tested. As in many other situations the law itself is fairly simple, but its interpretation and application to particular facts can be exceedingly complex. The use by an employee of legally prescribed drugs such as tranquillisers is unlikely to be sufficient grounds for dismissal. But it would depend upon the specific circumstances and the extent to which it could be demonstrated that by the use of such medication the individual was placing himself or others at undue risk. For example, the breach of a contractual obligation requiring an airline pilot to disclose all medication may, in certain circumstances, justify disciplinary action leading to dismissal.

In all cases of drug abuse in the workplace the employer

should seek expert legal advice, preferably at an early stage. Drug abuse is of growing importance to industry, not only because of its effects on production, absenteeism and accident rates, but also because of its legal and social implications. Chapter 11 outlines some basic strategies designed to combat the damaging effects of drug abuse in the workplace.

CHAPTER 8

STRESS – DEFINITION, CAUSES AND CONSEQUENCES

'It was the best of times, it was the worst of times, it was the age of wisdom, it was the age of foolishness, it was the epoch of belief, it was the epoch of incredulity. It was the season of light, it was the season of darkness, it was the spring of hope, it was the winter of despair. We had everything before us, we had nothing before us, we were all going directly to heaven, we were all going directly the other way.'

Charles Dickens *A Tale of Two Cities*

It would be hard to imagine a more apposite description of our own period of human history than the above, published in 1859. We live in astonishing times. The promise for the future is great, and yet the threat of extinction is real and immediate. In the next ten or 20 years we will not only determine the quality of life for ourselves and our children, but may well exert a decisive influence on whether, as a race, we shall survive at all.

The threat to our continued existence posed by the possibility of nuclear war has dominated our thinking and our attitudes for so long that it is difficult to imagine a world which is devoid of such a threat. There are now some grounds for optimism, but a world completely free of nuclear weapons is still a long way off.

And as one threat begins to recede slightly, another emerges. The threat of ecological disaster is more insidious but no less serious. We are rapidly approaching a global environmental crisis, which originates in our craving for technological, economic and institutional growth. The earth's natural resources, such as oil, coal, minerals, metal, forests, fish and oxygen have been, and continue to be, used at an alarming rate. As E.F. Schumacher says:

> An attitude to life which seeks fulfilment in the single-minded pursuit of wealth – in short, materialism – does not fit into this world, because it contains within itself no limiting principle, while the environment in which it is placed is strictly limited. Already, the environment is trying to tell us that certain stresses are becoming excessive.

And we in the West have been far more irresponsible and guilty in this sense, than have the peoples of the undeveloped countries. In the USA, for example, where excessive consumption and waste have become almost a way of life, we have a situation which 5 per cent of the world's population are now consuming about one third of its resources, with energy consumption per capita about twice that of most European countries.

But not only have we wasted our natural resources, we are now having to face up to the appalling mess that we have created in the process. The problems of toxic waste disposal, both nuclear and chemical; acid rain; the depletion of the rain forest in South America; the 'greenhouse' effect due to the burning of fossil fuels; the serious depletion of the earth's ozone layer due to chlorofluorocarbons (CFCs); the incredible traffic congestion and air pollution; the pollution of our rivers and our beaches; and because of the excessive use of chemical fertilisers, pesticides, hormones and various other chemical agents, food pollution is now a reality also.

If we asked ourselves why we do all this, we would probably be hard pressed to give an answer. Perhaps, somewhere in the back of our minds still lurks the idea that our pursuit of material wealth will ultimately lead to happiness. Unfortunately, the evidence would suggest otherwise. The late Erich Fromm, one of the most penetrating social analysts and psychologists of our time, wrote a remarkable book called *To Have or To Be.* In that book he discusses the myth of our happiness:

> . . . but even without theoretical analysis the observable data show most clearly that our kind of 'Pursuit of Happiness' does

not produce wellbeing. We are a society of notoriously unhappy people: lonely, anxious, depressed, destructive, dependant – people who are glad when we have killed the time we are trying so hard to save. Ours is the greatest social experiment ever made to solve the question of whether pleasure can be a satisfactory answer to the problem of human existence. For the first time in history the satisfaction of the pleasure drive is not only the privilege of a minority but is possible for more than half the population. In the industrialised countries the experiment has already answered the question in the negative.

At this point the reader may be wondering just what all this has to do with stress. It has everything to do with it. Because the essential ingredient in all forms of stress and anxiety is that they contain some element of *threat*. And the global threat is, whether we like it or not and whether we are conscious of it or not, the greatest threat that we face. Granted we are not conscious of these issues for most of the time, indeed life would be intolerable if we were. But we cannot avoid confronting them almost daily, in our newspapers and on television and radio, and they are bound to colour our view of the world and our hopes for the future. We may only be conscious of them for relatively short periods of time, but they constitute a background of anxiety which underlies everything we do.

Overlaid on the general anxiety are the day-to-day causes of stress which have their origins primarily in our working environment and in our personal relationships.

The organisation as stressor

Companies and corporate organisations are in a constant state of flux, of evolution, and because of this they can be a source of stress. The *rate* of change which characterises modern corporate life is particularly stressful. Mergers and takeovers, asset stripping and changes in leadership and senior management are going on all around us. We live with the threats, euphemistically talked about, of 'downsizing' (which means sacking people), 'decentralising' (which means sacking people at head office), 'delayering' (which means sacking selected members of management), and 'cost containment' and 'reducing overheads' (which may well amount to more sackings).

Alvin Toffler, author of *Future Shock*, defines change as 'the way in which the future invades our present'. To a large extent, our future has already arrived in the form of that most powerful catalyst for change, the microchip. The 1980s will undoubtedly be viewed in retrospect as the era in which computer technology came of age. It is the transition point between an industrial and technologically-based society to an information and knowledge-based society. Computers, microcomputers, word processors, laser printers, portable fax machines, and various software systems are regarded by many today as being normal and essential office equipment. But the revolution yet to come will make these machines look archaic. The key words of the next decade will be 'transmission' and 'integration'. The advent of world networking, satellite teleconferencing, computer conferencing, electronic mail, videotext and teletext, will allow us to distribute, receive and analyse data from an enormous number of sources, simultaneously. Our ability to store and retrieve information will take on a new dimension; we already have 12-inch optical discs which are capable of storing digital data equivalent to one million pages of full text on a single side.

These new computer technologies will provide us with an information system of astonishing power and complexity, which will have far reaching implications not only for corporate organisations, but for society as a whole. Those companies and organisations with the necessary skills and flexibility to adapt and innovate in this rapidly changing environment, will prosper. Those that do not, will simply atrophy and die.

According to Alvin Toffler, there is no technological reason to prevent 5 per cent of workers turning out all the material goods that we require. This implies a workforce in which as many as 90 per cent will being engaged in services, information and communication activities. The implications in retraining terms are obvious. Then there is the sheer amount of information that we are having to deal with: it has been claimed that our current knowledge base is doubling every ten years, and, in some of the sciences, every five years. Any wonder that we are all suffering with information overload, with what Saul Bellow has called 'the congestion of modern consciousness'. The obvious futility of even attempting to keep up with this flow of information in no way lessens its potential for anxiety and stress.

Personal stressors

All these stressors – global, socioeconomic and organisational – focus on the individual human being. Change is threatening to us because it demands that we ask certain questions of ourselves, questions about our ability to adapt and to learn new skills, and questions about our willingness to go with the change.

To all this we have to add the conflicts and pressures of our personal relationships, professional, social and family. We worry about our children, about money and about our jobs. We feel hurt and isolated when family and friends let us down. And we worry about getting old, about illness and about death. And so on.

In the discussion thus far I have tried to indicate that the potential sources of stressors in our everyday life derive from an enormous variety of sources, from global and socioeconomic forces at one end of the spectrum to family and personal relationships at the other.

In the face of all this it is perhaps not so surprising that some of us are suffering from stress.

Our capacity to cope

The reason why most of us are not submerged in a sea of despair is that we are not merely passive recipients in this process that we call living. We act on the world and on our environment, we influence things, we ourselves bring about change in our lives, and we exert some degree of control over what happens to us. We also have enormous resilience, which means that most of us, for most of the time, *are* coping. Moreoever, the need to deal with these external realities and threats is part and parcel of the normal process we call living, and it brings about its own sense of fulfilment and exhilaration. It would be a strange and insipid world if there was nothing for us to test ourselves against, and deeply enriching experiences can come out of adversity. That is not to say that we are all walking around with broad smiles on our faces all the time. Indeed, our attachment to life varies on a day-to-day basis. But for most of the time we have some sense of fulfilment, some sense of wonder, and a real sense of hope for the future.

What exactly is stress?

It is possible, as the preceding pages show, to talk about stress, its origins and causes, without saying what exactly stress *is*. It has been defined as follows:

> an adaptive response, mediated by individual characteristics and/or psychological processes, that is a consequence of any external action, situation or event that places special physical and/or psychological demands upon a person (Ivancevich and Matteson)

and

> the non-specific response of the body to any demand made upon it (Selye).

But perhaps a more practical and realistic definition would be as follows:

> when the problems presented by everyday life exceed your ability to cope with them, you feel stressed.

Of course this still doesn't tell us exactly what stress *is*, it merely helps to define a situation in which we may experience it. Trying to define the word 'stress' is rather like trying to define the word 'blue'. We all know what the colour blue looks like, we all recognise it when we see it, but we could not in any real sense define it. Stress can mean so many different things, depending upon the particular setting and the individual concerned, that definition is difficult and probably unnecessary. We shall, therefore, proceed on the assumption that the word is understood without definition.

However, it is extremely important to recognise that stress is not invariably a bad thing. Although the harmful effects have received a lot of attention over the last few years, the fact remains that some of the most rewarding moments in our lives are moments of great stress. Stress is important in spurring us on to achieve our ambitions and so adds immeasurably to the quality of our lives. Indeed, many people who have little stimulation from their work feel compelled to go out and look for sources of stress, to find interest and excitement. This is why people enjoy parachute jumping, hang gliding, racing cars or running marathons. It is only when stress becomes unduly severe or prolonged that problems can arise and stress can then become a very serious negative force.

How prevalent are stress-related disorders?

It is widely believed that stress-related disorders are now reaching epidemic proportions. Some suggest that as many as one third of our workforce are currently suffering from stress-related disorders, and that this results in a loss of about 30 million working days per year at a cost of anything between

£3 and £10 billion. But the fact is that we have no real idea how common stress-related disorders are, let alone how much they are costing us. As we have said, stress has so many forms healthy and unhealthy, and varies so much in degree and effect, and affects individuals in such different ways that its extent in the population is impossible to quantify. At what point do you decide that a normal stress reaction becomes damaging and, more importantly, how do you go about measuring it? If you cannot adequately define it, then you cannot measure it. If you cannot measure it, you cannot estimate its prevalence, let alone infer its economic consequences.

This is not, of course, to say that stress is unimportant either from the point of view of human suffering or from the point of view of loss of financial performance. It is just being sceptical about current estimates of the scale of the problem. We should not forget that stress is an essential ingredient of our working lives and too little of it can be just as damaging as too much. Moreover, stress should not be used in the glib way that is fashionable today. Twenty years ago, people were lazy, inefficient or frankly incompetent. Nowadays, they have 'stress-related nervous debility'.

Who gets stress-related problems?

The words 'executive' and 'stress', tend to go together like 'sex' and 'violence'. The implication is that for some reason stress-related problems are more common in executive and senior managerial groups than in other people. It is difficult to understand the origins of this mythology; perhaps executives like to propagate the idea as proof of the great personal sacrifice they make for their particular organisation. Whatever the reasons, the notion that executives have the monopoly on stress is patent nonsense. Nevertheless, many people appear to believe it. In a 1987 MORI survey, the personnel directors of 112 financial organisations were asked a number of questions relating to health issues in their organisations. More specifically, they were given a list of possible health problems and asked to assess how important each one was, on a 0 to 5 scale. So far as the management and white collar staff were concerned, the personnel directors felt that the most important health issue was stress. This indicates the strength of the perception that stress is a significant factor in leading to ill health. The personnel directors believed that stress was not such an important issue among the blue collar employees and that its manifestations in terms of symptoms differed between blue collar and white collar workers.

It is reasonable to suspect that this impression was not based upon any hard evidence, but was actually an expression of the generally accepted view that those employed in executive positions, are under greater pressure and are therefore much more liable to be subjected to the adverse effects of stress. One may also suspect that if the blue collar workers had been invited to participate in the survey, their vision of reality would have been rather different! It does not seem logical to say that people in specific occupations or at specific levels within an organisation are more prone to the effects of stress than others, because stress is such an intensely personal issue. Stress can affect all of us, in all walks of life and at whatever level. What about the stress in the single parent family? What about the stress felt by the young mother with two children who pester her all day long, and then cry all night? And what about the stress which arises as a consequence of being unemployed? It has been shown very clearly how unemployment is related to poor health (both physical and mental), premature disability, suicide and death.

There are some factors which may make some of us more vulnerable at certain times. An accident or illness may well make us more prone to the adverse effects of stress, and may, in some cases, lead to true depressive illness. The amount of support which we receive from family and friends, and our ability to communicate and articulate our anxieties, are important factors too. Personality type has attracted a good deal of attention in recent years because of growing evidence that our intrinsic personality may not only determine our level of exposure to stress, but may also place us at particular risk from its effects. (See p 220.)

What happens during an acute stress reaction?

In the language of stress theory, *stressors* are outside demands that trigger an adaptive *stress response* within the individual. It is important to emphasise that these stressors may be positive and pleasant, or negative and unpleasant. The essential point is that the stress response represents a relationship, an interaction, between the individual and the stressor. In stress theory, the word to describe this is *transactional,* which indicates that stress is influenced both by the individual and by the environment. In other words it is not simply an event or series of events that causes stress, but also the way a person views such events. It is the individual's appraisal and perception of a particular event, rather than the event itself, which is crucial. After all, what is threatening and stressful for one person is not necessarily so far the next. A graphic example of this is the range of reactions within a group of people who have been held hostage by a group of terrorists. At one end of the spectrum is the experience of hysteria and paralysing fear, and at the other end there are individuals who remain completely cool and controlled in spite of the terrible circumstances in which they find themselves. Moreover, it is extremely difficult to predict how any individual will react to any circumstance at any point in time. But whatever the reaction is, it is fundamentally an individual one, and is brought about by a dynamic relationship, a transaction, between the individual and his or her environment.

It is important to understand that a stress response may be entirely appropriate. If a hooded gunman held a revolver to your head, you would experience an acute stress reaction – it would be abnormal if you did not. If you worked in an organisation which had a reputation for sacking people and where there was therefore a constant threat of dismissal, you would feel anxious and stressed. Again, this is a normal and appropriate response. Stress reactions, therefore, are normal responses to abnormal situations. Your perception of that abnormal situation is exceedingly important, and may even be critical to your survival. It has been said that 'if you keep your head when all about you are losing theirs, it's just possible that you haven't grasped the situation'.

When we do feel a real threat, what actually happens to the body? The highly complex series of biochemical and psychological reactions have been called the 'fight or flight' response. It is genetically programmed into each of us and links us to our prehistoric ancestors. A rapid outpouring of the 'stress hormones' (adrenalin and noradrenalin) produces an increased state of arousal; this is quickly followed by a number of other changes, including an increased pulse rate, increased blood pressure, dilation of the pupils and an increase in blood fats and blood sugar. (See Figure 8.1.)

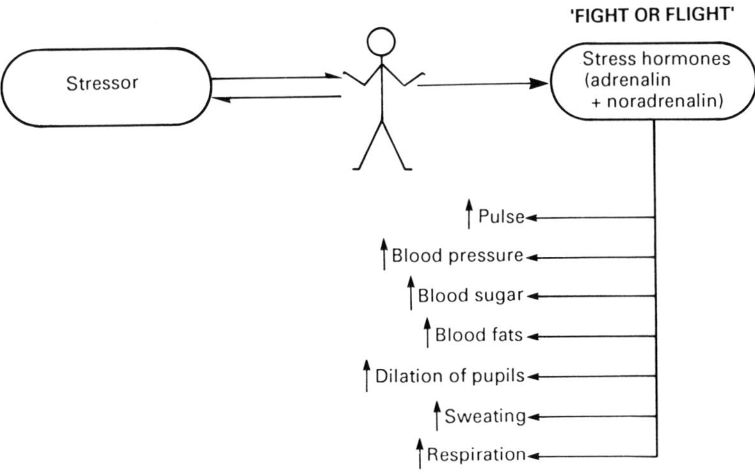

Fig. 8.1 *The human 'fight or flight' response*

Why is the stress reaction harmful?

If the reaction described above is a normal response to an abnormal situation, why is it potentially harmful? The reason is that it is a lifesaving mechanism, designed to be activated occasionally. If the stress response is required continually or repeatedly, putting the body in a high state of arousal for long periods, the mechanism can become life threatening. The body begins to show signs of wear and tear, and if the stress transaction goes on long enough it will eventually break down completely.

Type A and Type B personality

It has been recognised for many years that personality and behavioural traits may affect the outcome of disease. The Type A and Type B designations originate from a study done in the late 1950s on groups of patients with heart disease. The researchers found two major forms of behavioural pattern in the subjects, which they designated Type A and Type B personalities.

The Type A pattern is characterised by intense ambition and competitive drive, constant preoccupation with deadlines, extreme impatience with delay, anger and hostility. We all know this sort of person. He is the one who is always finishing sentences for you, and during the course of conversation is forever looking at his watch as if he has to go somewhere else, to talk to someone else.

The importance of the Type A behaviour pattern is that it identifies those who are much more likely to develop coronary heart disease. A number of studies since have supported this association, including two large epidemiological studies – the Western Collaborative Group Study and the Framingham Heart Study, both carried out in the United States. The former, carried out over an eight-and-a-half-year period, showed that Type A individuals had about twice as many heart attacks as did Type B individuals.

The hostility and anger components of the Type A personality are particularly interesting. The hostility is 'free floating', in other words it is not directed at anything specific but at everything in general. There is often no rational or reasonable explanation for this attitude. Although we use the term 'hostility', it would perhaps be more accurate to describe it as an attitude of cynicism and lack of trust in other people. The Type A person assumes that people will be unpleasant, nasty and vindictive until given hard evidence to the contrary. This lack of trust in others may be responsible for some other features of the Type A behaviour pattern. For instance, if one cannot trust others, then it is all the more important always to be in *control.*

It appears that inwardly directed anger – 'bottled-up anger' – is much more damaging to the health than outwardly directed anger. In other words, the individual who says 'even when I am very angry, most people around me wouldn't know about it', is much more at risk than the individual who says, 'when I get angry you will know all about it'.

It is important to recognise that although intense ambition and competitive drive are features of the Type A personality, they are not found exclusively in high flyers. Indeed, Type A personalities very often do not achieve promotion to the highest levels in an organisation, simply because they are not suited to teamwork and cooperation with other people. Also, they very often lack the application to focus on one particular issue, preferring instead to jump from one project to another, and spreading themselves so thinly in the process, that they barely make progress with any of them. Type A individuals can be found at all levels in an organisation and not only, as is commonly believed, in senior management.

The Type B individual is *not* simply the converse of Type A: lazy, little ambition, lack of competitive drive. The Type B personality may well be strongly achievement-orientated and ambitious but he tends to be more consistent and to have greater concentration and application than his Type A counterpart. He is the sort of person who will keep working away quietly and methodically, building in the process, much more complete relations with others. He is more organised, more participative, more cooperative and is usually an effective team member. He is often much more successful at achieving the highest managerial and executive positions than his Type A counterpart.

Only about 10 per cent of the population can be regarded as being purely Type A; a further 10 per cent are Type B. This means, as one might expect, that the vast majority of us are a mixture of both personality types, with a tendency for one or the other to dominate. Whether we are mainly Type A or not, most of us get impatient now and again. (Some research carried out by a large American fast food organisation reckons that it only has about 40 seconds to serve you, before you start to show evidence of impatience.)

Why does the Type A personality appear to be more prone to heart disease? The mechanism is not yet fully understood, but it appears to be connected with the levels of stress hormones such as adrenalin, nordrenalin and cortisol. Type A individuals seem to exhibit much brisker hormone responses in stressful situations than Type B individuals. These stress hormones may be capable, in some circumstances, of causing damage to the lining of the arteries and contributing to the 'furring up' process which is the hallmark of coronary heart disease.

What are the symptoms and signs of excess stress?

We need to consider these in terms of acute and chronic effects. The acute 'fight and flight' reaction was mentioned earlier. Symptoms include sweating, rapid heart rate, dilated pupils and increased blood pressure. For short periods high level of stress may be associated with increased productivity, greater personal achievement and even a sense of exhilaration. However, if the stress continues for long periods, productivity and efficiency begin to fall off rapidly and signs and symptoms of overload may begin to appear. (See Figure 8.2.) Symptoms vary greatly, and occur in different combinations in different people.

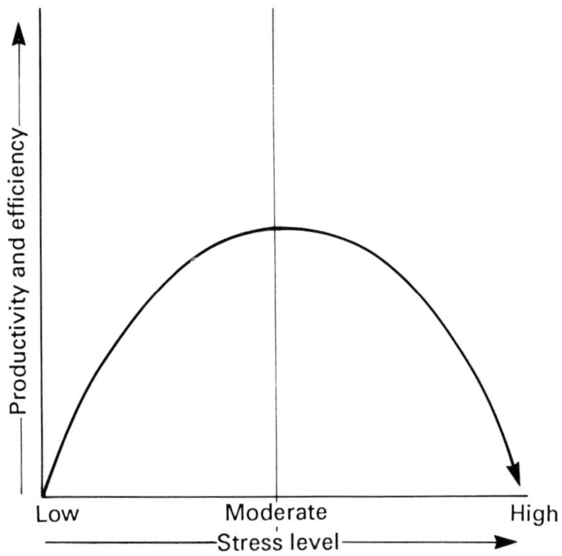

Fig. 8.2 *'Human function curve' – low to moderate degrees of stress increase productivity and efficiency, but sustained high levels of stress lead to breakdown and a marked reduction in efficiency*

The more important are as follows:

- Tiredness and lethargy commonly associated with changes in sleeping patterns, particularly early morning wakening and difficulty in getting off to sleep.
- Loss of interest, apathy, boredom and depression
- A feeling of anxiety, tension and fear
- Physical symptoms: sweating, palpitations, indigestion, disturbed bowel function, breathlessness and headaches
- Rapid and unexplained mood changes, with aggressiveness and outbursts
- Poor concentration and memory
- Reduced sex drive

- Indecision
- Increased alcohol, cigarette or drug consumption.

As often as not, an individual will deny any such symptoms exist even when the signs of undue stress are glaringly obvious to family, friends and colleagues. These signs may include the following:

- Reduced performance
- Increased absenteeism
- Excess consumption of alcohol and cigarettes
- Extreme irritability and rapid mood changes
- Emotional instability
- Indecisiveness
- A look of extreme fatigue, nervousness and/or depression.

Attempts to discuss any of these issues will often be met with rejection, strong denial or even an aggressive outburst.

What are the long-term consequences?

In the long term, the individual who suffers chronic, unremitting stress, may be placing himself at risk of serious disease or even death. In recent years there has been increasing recognition of the relationship between stress and a variety of diseases. Experimental evidence suggests that under prolonged stress the body's ability to fend off disease may be seriously impaired because its immune system is damaged. Part of this evidence is the discovery that the immune function of widowers is depressed for a year after the death of their wife. Most of us notice that we are more prone to get cold sores and other minor infections during periods of increased stress.

Nowadays we accept that no rigid distinction can be made between the mind and the body, and that neither illness nor health can be defined in purely physical terms. We all know from personal experience that the way we feel emotionally and psychologically has an enormous bearing upon how we feel physically. (It is perhaps surprising that medical practice has not adapted to reflect this unity.) It affects not only our susceptibility to a wide variety of diseases, but also our capacity to overcome them once they have occurred.

We have discussed the relationship between stress, personality and heart disease, but prolonged stress can also lead to other serious forms of ill-health: alcoholism and other forms of drug dependency, depression and suicide may also result from prolonged stress.

To help them tolerate life under the system, the citizens of Aldous Huxley's *Brave New World* were given a regular ration of 'soma' which, apparently, possessed 'all the advantages of Christianity and alcohol, and none of the defects'. Unfortunately, our own medications are not without defects. We now know that drugs like Valium, Librium and other forms of tranquilliser cause serious problems of addiction and withdrawal.

Finally, there may be a relationhsip between psychological factors and cancer. The idea that there could be a link of some sort has ancient origins, beginning, according to medical historians, in the second century when Galen observed that cancer occurred more commonly in 'melancholic' women that those of 'sanguine' temperament. Eighteenth- and nineteenth-century medical literature contains numerous references to emotional distress preceding the onset of cancer. Modern scientific evidence of such a relationship, however, is contradictory. Some studies have confirmed an association, while others have strongly rejected it. There has been a lot of clinical observation over the years, of stressful experiences apparently preceding the onset of cancer, but the most we can say at the moment is that there is a *possibility* of an association.

Stronger evidence does, however, exist that stress can influence the course of cancer once it has occurred and does therefore affect survival rates. Research reported recently in the *British Medical Journal,* carried out at Guy's Hospital, shows that women with breast cancer are nine times more likely to have a recurrence if they experience a stressful life event. Research at the Royal Marsden Hospital, has also shown that breast cancer patients who show fighting spirit, tend to survive longer; current patients at that hospital are offered psychological therapy in an effort to transform negative thoughts into 'positive coping responses'.

Stress – whose responsibility is it?

We have seen that sources of stress are manifold, which often makes it difficult or impossible to be sure of the cause in a particular case. An individual in an organisation clearly has a responsibility to do whatever he can to prevent the stress factors in his personal life from impingeing on his performance at work. Equally, the organisation has a responsibility to ensure that undue stressors are not placed upon the individual.

There is here a potential source of conflict. Organisations may blame individuals for not controlling stress sources within their own lives, and individuals may blame their employers for creating a stressful working environment. In California the courts have been awarding employees damages for stress-induced illness, where employers are judged to have done nothing to minimise the effects of stress within the workplace.

Obviously, responsibility is divided between the individual and the organisation. Organisations can and should provide a supportive environment that encourages employees to take more responsibility for their own health and wellbeing. Equally, the individual has a responsibility to the organisation to control those factors in his own life that could have an adverse effect upon his performance at work. Co-operation, understanding and tolerance are needed on both sides.

What can we do about stress?

In the MORI poll mentioned above, it was interesting that stress was recognised as an important health issue, but few organisations believed that they could do anything constructive to limit its effects. This chapter has emphasised the 'transactional' – dynamic – nature of stress. We are not passive recipients of stressful events. On the contrary, we *act* upon our environment, we change our circumstances, we influence events. So there is a great deal that can be done to manage the potentially damaging effects of stress, both at an individual and at an organisational level. Chapter 13 describes some appropriate strategies.

PART 3

INTRODUCTION

This section of the book discusses some of the ways in which lifestyle factors interact together and contribute to diseases, particularly cardiovascular disease and cancer. First it is necessary to emphasise a few points:

1. Lifestyle-related risk factors do not account for *all* cases of these various diseases.

2. It follows that changing these risk factors will not *prevent* all cases of these diseases.

3. The relative contribution of each of these factors to each disease varies considerably. For example, stopping cigarette smoking would prevent more than 95 per cent of all cases of lung cancer, whereas it would have a much smaller impact on other types of cancer.

4. Calculating the level of risk when more than one risk factor for a disease is present is not simple. For example, the presence of cigarette smoking and high blood cholesterol does not double the risk of developing heart disease, but increases it many times.

5. Risk factors are strongly interrelated. Hypertension is a major risk factor for heart disease, but hypertension is itself strongly associated with obesity, a high salt intake, and excess alcohol. Again, there is a strong association between a high fat diet and various types of cancer, but a high fat diet is also associated with high levels of blood cholesterol, which is in turn a major risk factor for heart disease.

Since most of the diseases which we shall be discussing involve a multiplicity of risk factors, it is clear that close attention must be paid to *all* of them to keep overall risk at a minimum. Taking regular exercise will obviously help to reduce the risk of a heart attack, but it will have considerably less benefit if you are a cigarette smoker and have a high blood cholesterol level.

CHAPTER 9

THE ROLE OF HEALTH SCREENING IN PREVENTION

Large numbers of senior executives and managers in this country undergo an annual medical examination or check-up, carried out either by a physician appointed by the company, or by one of the growing number of commercial organisations offering a variety of health screening services. Such examinations are usually fairly standard in terms of their core content, but in recent years the *range* of investigations on offer has increased considerably, so that specialised fitness testing, nutritional analysis and even blood tests for the early identification of certain forms of cancer are now becoming available. In the near future these tests are likely to be superseded by even more sophisticated and refined investigations which will greatly facilitate the early identification of diease.

Health screening is likely to become even more important and it will not be easy for the non-specialist – senior managers, personnel directors and so on – to decide what sort of programme would suit their own particular organisation. It is therefore pertinent to look at the medical value of screening programmes and to see whether the potential benefits to the organisation can justify the very significant expense involved.

Organisational attitudes

A recent study on employee attitudes conducted by the Henley Management College indicated that good healthcare provision ranked second only to opportunities for further education in the employees priority ratings. Given this fact, it is surprising that so little thought often goes into the huge investment in these services. There is usually a vague idea that providing health screening is a good thing, but most companies would be very hard pressed to say exactly what they expected the programme to deliver, much less have any mechanism of quantifying it. Perhaps one of the main reasons for this is that health screening has been in the past regarded as nothing more than a company perk – a means of enhancing a particular remuneration package.

Most organisations in this country still only provide health screening to senior management and executives – usually no more than the top 10 per cent in the company. This practice does not begin to make any sense, either from a medical or an economic perspective. First, the diseases, and the risk factors one is seeking to prevent or to modify, are actually more prevalent in the lower social class groups. Second, all relevant long-term studies have demonstrated that financial benefit derives from investment in such programmes when the *whole* of the workforce, rather than a small minority, is involved.

The main reason why many companies offer health screening in the first place is because it is regarded as an important perk; it is a way of saying to key people in the company: 'You are very important to us, and we are concerned about your long-term health and wellbeing'. Unfortunately, the equally clear message thereby sent to the rest of the workforce is: 'You are *not* important to us and we don't really care about your long-term health prospects'.

Organisations need to differentiate between their treatment of certain sections of the workforce in order to create the incentive to achieve. But to offer the most comprehensive health programmes to the key executives at the expense of everyone else is not only medically unsound and economically unrealistic, but is, in my opinion, morally and ethically questionable. If there is a choice between something more basic for everyone and something very sophisticated for a small minority, then there is no question as to how the organisation should utilise its resources. Of course, once a basic programme is provided for everyone, there is no reason why an organisation should not then provide a more detailed and intensive screening examination for key individuals within senior management.

The annual medical examination itself has, in the past, often amounted to little more than a comfortable chat with the company doctor who tells you to smoke and drink a bit less and lose a few more pounds, then pats you on the head and tells you that you are basically in good shape. The point of this process is mainly to reassure rather than to educate or motivate.

All of this has changed dramatically during the past five years. With a higher level of awareness about disease risk factors, and the costs of absenteeism and sickness, decision makers in companies are asking much more searching questions about what they are actually getting from their investment in health screening programmes.

What are the reasons for health screening?

Health screening is a set of tests and examinations applied to an apparently healthy (asymptomatic) population, with two specific aims in mind:

- To identify disease which may already be present, but which may not yet have produced symptoms (for example, hypertension)
- To identify and quantify various risk factors which may predispose to the development of disease in the future.

Identifying disease as early as possible makes a complete recovery more likely than if the disease is only discovered when the symptoms become evident to the patient. The idea that the early detection of a disease *invariably* gives a much higher chance of cure is an attractive one, but it is not easy to prove.

The identification of disease risk factors such as high blood cholesterol, obesity and cigarette smoking in an individual, and then modifying them, leads to a much lower probability of developing disease in the future. Thus the aim of the programme is not only to identify these factors, but to educate and motivate the individual to make appropriate behavioural changes.

On a population basis there is good evidence that the modification of risk factors can lead to a significant decrease in disease. However, it is important to note that this cannot provide any guarantee in the *individual* case. We know, for instance, that reducing blood cholesterol in a given population, will reduce the overall incidence of deaths from heart disease. Therefore, on a population basis, reducing blood cholesterol levels appears to be an appropriate strategy. The problem is that it is impossible to be certain that reducing blood cholesterol in a particular *individual* will reduce his risk of heart disease. The best that we can say is that, on the basis of probability, it is a recommended strategy.

The requirements of a screening programme

The World Health Organisation has drawn up guidelines for screening which state that the disease being screened for must:

1. Be common in the population being screened
2. Be serious
3. Have a long latent phase (that is, take a long time to develop)
4. Have an eventual outcome which is beneficially influenced by early diagnosis
5. Have a treatment which is both available and acceptable to the patient.

The screening test which is used to identify that disease must be:

1. Positive in the majority of patients who have the disease
2. Negative in the majority of patients who do *not* have the disease
3. Economical in terms of medical resources
4. Acceptable to the population
5. Do no harm to symptom-free people.

In some of the diseases screened for nowadays it is not always possible to fulfil all of these criteria completely. For example, screening for colorectal cancer fulfils most of these principles, but it is not possible at the moment to guarantee that early diagnosis will *necessarily* lead to better survival and recovery prospects. Logic certainly suggests that it will and major studies are currently underway, but there is no certainty about this. In any event, the above points are meant only to be guidelines rather than a rigid set of rules.

Knowledge/attitudes/behaviour

Most organisations who provide health screening do so on an annual or biennial basis, usually according to the recommendation of the designers of the programme. This periodic check-up is very often the beginning and the end of the matter for the individuals concerned.

The value of this sort of programme in producing real changes in risk factors is questionable. Very few people make any real and lasting changes in their behaviour as a result of this type of health screening. Companies are often surprised to find that the number of their employees who continue to smoke or who take little or no activity remains the same, despite repeated exposure to health screening programmes. The programme is operated in the somewhat naive hope that giving people information about their current health status is, in itself, sufficient stimulus for them to initiate remedial action. But in most cases, this simply does not occur. Many people make changes for a while and then revert to their old ways again. Additional information or education is needed to change attitudes enough to produce more healthy behaviour patterns.

For example, a cigarette smoker knows that cigarette smoking may cause lung cancer, but his attitude – 'Well lots of people smoke cigarettes and don't get lung cancer, so I will probably be one of the lucky ones' – remains, and does not lead him to change his behaviour. But if he is given the knowledge that his smoking may not only cause lung cancer in himself but may also harm his young child, his attitude may change. He may now say: 'smoking may or may not cause me harm, but I cannot bear the thought of my own habits damaging the health of my child'. This attitude is likely to be much more conducive to a successful attempt to stop smoking.

Health screening programmes must, if they are to be really effective, be supported by a continuous programme of health maintenance and information to support individuals in their efforts to change their behaviour.

The content of screening programmes

As we have said, the various health screening programmes currently available are usually based around a central core of investigations and tests. Companies are also able to choose more specific forms of screening, such as heart disease screening for men or 'well woman' cancer screening. The ability to select specific screening services can save costs. Figure 9.1 shows the content of most screening programmes currently available.

Programmes like the one in Table 9.1 are sometimes termed 'multiphasic'; that is they screen for a number of diseases simultaneously. At first sight, you might think that the more tests are included, the more effective the programme is likely to be. In fact more does not mean better, and while many of the tests shown in Figure 9.1 are essential, others are much less valuable or (in some cases) irrelevant.

Table 9.1 *Content of screening programmes*

- Health questionnaire and medical history
- Height/weight, Body Mass Index (BMI)
- Body skin-fold percentage
- Blood pressure (BP)
- Lung function test
- Hearing test
- Sight test (including glaucoma)
- Chest x-ray, abdominal x-ray
- Urine test
- Resting ECG (electrocardiogram)
- Stool test (occult blood)
- Blood tests for liver, kidney and metabolic disease, including measurement of blood fats (e.g. cholesterol + HDL)
- Consultation with physician and thorough physical examination

For women only

- Breast examination
- Instruction in breast self-examination
- Breast x-ray (mammography)
- Pelvic examination
- Cervical smear

. . . .

Additional tests may also be available and include:

- Fitness testing and assessment
- Nutritional/dietary analysis
- Ovarian cancer screening
- Functional/diagnostic graded exercise stress testing (FDGXT)

The customer may believe that a chest x-ray and a resting ECG are essential components of any proper screening exercise, but this is no reason to include them in the programme. Medical and scientific evidence alone should dictate the content of screening programmes. Professional screening services should be able to demonstrate a commitment to providing high quality information and advice to their clients and be able to justify what they recommend on medical and scientific grounds.

Some organisations may simply want to provide comprehensive health screening for a part or the whole of its workforce, irrespective of the cost involved. They may have reached the conclusion that the high perceived value of this type of service more than compensates for the investment involved. For these organisations, the fact that some of the screening tests are much more useful than others, or that some of them are actually of very little use at all, is of no consequence. It is the value the individual employee places upon the service which is the overriding factor.

For most companies purchasing screening services, however, cost is a big consideration, and the primary objective of the exercise is to provide a screening programme which will have as much impact upon the health of the workforce as possible, given the resources available.

It is therefore important to take a critical look at the content of the screening programme under consideration. For example, a standard chest x-ray is usually included in the more comprehensive ones, although in an apparently healthy population, the 'yield' (that is the number of abnormal chest x-rays) is exceedingly small. Most authorities nowadays regard a routine chest x-ray in a healthy population as being a waste of time and resources. The very remote possibility that something will be picked up, is not thought to justify the expense involved. Would it not be better to use the money available to finance investigations with a much higher yield, and to provide a continuing programme of risk management and health education for everyone?

The following components of the programme in Figure 9.1 are ones which can be regarded as non-essential, and can justifiably be omitted to save on costs.

- *Lung function test.* Lung function testing is useful where there is an occupational exposure to dust or other irritants, or for individuals with a relevant family history of chest disease, for example asthma. It is sometimes useful as a baseline test in smokers to indicate the effects of cigarette smoking on the lungs.

  In a non-smoking healthy population, with no obvious occupational exposure, the test is almost invariably normal.
- *Hearing test.* Again, this is obviously required where there is or has been an occupational exposure to noise. Otherwise, testing an asymptomatic population has limited value.

- *Sight test (including glaucoma).* Screening for visual defects in an apparently healthy population has a low yield, and sophisticated eyesight testing is, of course, already available from qualified opticians. Glaucoma is a condition which, if left untreated, can lead to blindness. Even though only a relatively small percentage of the population is at risk, provided the appropriate equipment is available, it is worthwhile screening for. But again, this is a service which can be carried out by an optician and so is not an essential part of a screening programme.
- *Chest x-ray, abdominal x-ray.* Again, in the absence of symptoms or relevant history it is difficult to justify either of these investigations. The yield is extremely low, and we should not ignore the fact that the individual is being unnecessarily exposed to x-rays, albeit an extremely small dose. The era of the mass chest x-ray is over; it was originally introduced because of the high prevalence of pulmonary tuberculosis. Some screening organisations still recommend the use of a baseline chest x-ray, particularly in smokers. But if a person is a smoker you will tell him to stop smoking; if he is a smoker and he has a chest x-ray which is normal, you will still tell him to stop smoking; if he is a smoker and the chest x-ray reveals a cancer on his lung, his outlook is extremely poor no matter what you do. So even for cigarette smokers, the chest x-ray is largely a waste of time and money.

  Situations in which the chest x-ray is appropriate are in occupations involving asbestos or other hazardous substances, when there is a history of chest disease, or when the individual has relevant symptoms. It may also be important in groups with a high risk of pulmonary TB or in patients with deficient immune systems, like AIDS cases, where the risk of chest infections is high.

- *Resting ECG (electrocardiogram).* In the absence of a family history or symptoms, a resting ECG is of extremely limited value. A normal resting ECG does not exclude the presence of severe coronary heart disease, so it is not a reliable basis upon which to reassure the patient of his fitness. Some would argue that the resting ECG provides a means of comparison on an annual basis and is therefore a reliable indicator of any changes that have occurred. The difficulty of this argument is that if the patient has no symptoms, it is by no means clear what ought to be done if changes are detected.
- *Blood tests.* Routine bood tests provide a useful means of identifying anaemia, alcohol abuse, kidney disease, liver disease and so on. They may also reveal that the subject is at a particularly high risk of developing heart disease. There is a range of blood tests that can be carried out, many of which are almost invariably normal in a healthy population. However, since most laboratories now carry out comprehensive 'profiles' on a fully automated basis, it is easier and most cost effective to have *all* the tests done rather than requesting certain specific ones.
- *Consultation with physician and thorough physical examination.* For a comprehensive physical examination, a doctor is obviously essential. But simpler screening programmes, involving large numbers of people can be carried out by an appropriately trained nurse. This allows an effective programme to be delivered but with considerable cost savings. For example, a coronary heart disease screening profile which involves a brief questionnaire, some simple measurements such as blood pressure and body weight and the taking of a blood specimen, does not require a doctor to be present, although a clinician obviously has to scrutinise the results.

Interpretation and communication of results

Whatever screening programme is used, it is obviously important that the results of the various tests should be interpreted by an appropriately qualified person and communicated clearly to the individual concerned. The information should be presented in such a way as not only to inform but also to motivate the individual to make the changes in his diet, exercise habits or drinking habits and so on, which have been recommended. The quality of the health information literature which is provided as a support to the screening process itself, is critical.

False positives and false negatives

Any screening test produces a certain number of false positive and false negative results.

A false positive result is a situation in which an individual has been told that he *has* a certain disease, when in fact he does *not*.

A false negative result is a situation in which an individual has been reassured as to the *absence* of a certain disease, when in fact the disease is *present*.

In order to keep the number of false positive and false negative tests to a minimum, close attention must be paid to an important principle known as Bayes theorum. In greatly simplified terms, this states that the value of any screening test is related to the prevalence of the disease in the population to which the test is being applied – the greater the amount of the disease in the population, the more valuable the test, and vice versa. For example, a screening test for heart disease in a population of women under the age of 25, would have very poor predictive value because heart disease in this particular group is uncommon. If the test were applied it would create a very large number of false positive tests. If, on the other hand, the test were applied to a population of males above the age of 55, the test would have a high predictive value because the prevalence of the disease in this group is relatively high.

When applying screening tests to any group of individuals, it is extremely important to have some information as to the prevalence of the disease being screened for and the accuracy expected from the test. 'Blanket' screening tests for diseases which have a very low prevalence in the population being studied, only create problems with the interpretation of results.

Designing a screening programme

The first stage in designing a screening programme is to establish four basic facts:

- How many people will be offered the service?
- What is the age distribution of this group?
- What is the relative proportion of male and female employees?
- What budget is available?

If you are in a position where there are no financial limitations and you wish to make multiphasic screening available to every member of the workforce, irrespective of age or sex, then the solution for you is quite straightforward.

If you need to be more selective, you should concentrate your resources on screening for those diseases which, depending upon the age and sex distribution of the workforce, are likely to be the most common. In basic terms, the emphasis when screening males should be on the detection and prevention of cardiovascular disease, while in women the main focus should be on the prevention and early detection of various types of cancer. Of course, heart disease affects women, and cancer affects men, and in an ideal world perhaps we would screen everyone for everything. But with limited resources we need to direct screening tests at those diseases which are most prevalent in the populations we are considering. Once this has been done, if resources permit, we can consider screening for less common diseases.

There are three simple tests which should be made available to *everyone*:

- *Blood pressure.* High blood pressure is a major risk factor for heart disease and strokes, and yet it often produces no symptoms whatever. The only way of knowing whether an individual has raised blood pressure or not, is to measure it. This is a simple and inexpensive test which takes only a few moments to perform. Identification of high risk groups would mean earlier and more effective treatment.
- *Blood cholesterol.* An elevated blood cholesterol level is a significant risk factor for cardiovascular disease. Early intervention would allow high risk individuals to make appropriate modifications in their diet and, if necessary, to be prescribed appropriate medication.
- *Urine test.* A simple urine dipstick test costs only a few pence to perform but can be effective in identfying a variety of conditions, including diabetes.

These three simple tests alone would represent a very worthwhile screening programme, particularly if they were supplemented with general information regarding diet, exercise, cigarette smoking and other risk factors.

Risk stratification

One useful approach is to use a simple screening test for the whole workforce, and, on the basis of its results, stratify them according to low, moderate, or high risk. In screening for heart disease, for example, the measurement of blood pressure, blood cholesterol, and Body Mass Index, together with information regarding family history of heart disease, exercise habits and so on, can be used to produce an overall risk score in any indvidual case. Having identified a group of high risk individuals, they can be specifically targetted with health information advice, more sophisticated investigations where necessary and closer follow-up and observation. (See Table 9.2.)

Table 9.2 *Three stages in risk stratification*

Stage 1
Baseline investigations and questionnaire for whole group

Stage 2
Interpret data and stratify according to low, moderate, or high risk

Stage 3
Target high risk groups with:
- Further tests if recommended
- Detailed health education advice
- Close follow-up and observation

This model is a sort of halfway house between a very basic programme on the one hand and an expensive, sophisticated multiphasic programme on the other. Its great advantage is that it allows the employer to direct resources towards those individuals who are most vulnerable, while at the same time providing *everyone* with an opportunity to participate in the screening programme and to benefit from the health promotion activities.

The four main steps in prevention

We have seen that health screening is an important part of prevention, but it is by no means the whole story. The process requires good interpretation and communication of results, careful explanation to the individual of what he needs to do to lower any risks, and adequate follow-up and support. (See Figure 9.1.)

At the time of the initial assessment, the presence of certain symptoms may require immediate referral to the employee's general practitioner. If, for instance, blood pressure was found to be very high, immediate referral to the GP would be required. If, on the other hand, blood pressure was found to be borderline, the next stage of the process is to communicate this fact to the patient and explain exactly what high blood pressure is and why it is essential to control it. The third

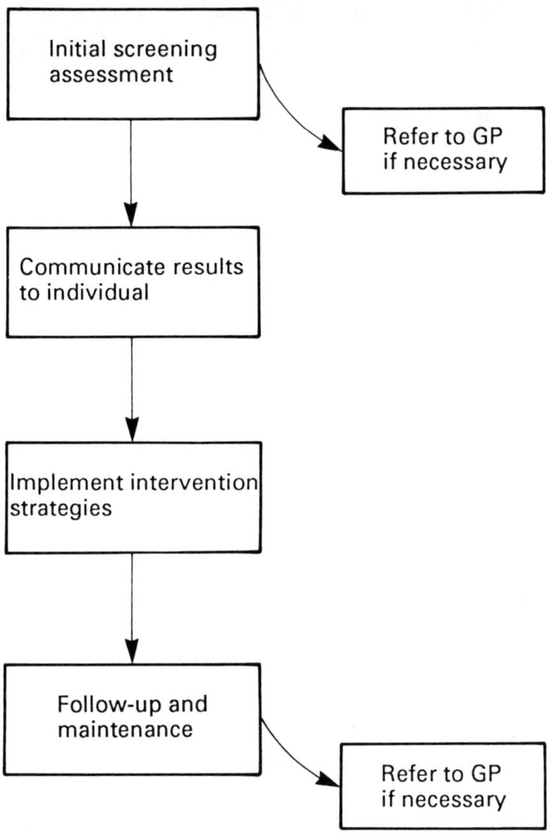

Fig. 9.1 *Four stages in prevention*

stage, intervention, would in this case, involve losing weight, reducing salt intake, taking moderate exercise and reducing alcohol consumption. The fourth stage would involve checking the blood pressure again after, say, three months. If it was still found to be high, then referral to the patient's doctor would be appropriate. If, as a result of the intervention, the blood pressure was within normal limits, the correct management would be to continue to reinforce the health education message and arrange to have the blood pressure checked at intervals. (See Figure 9.2.)

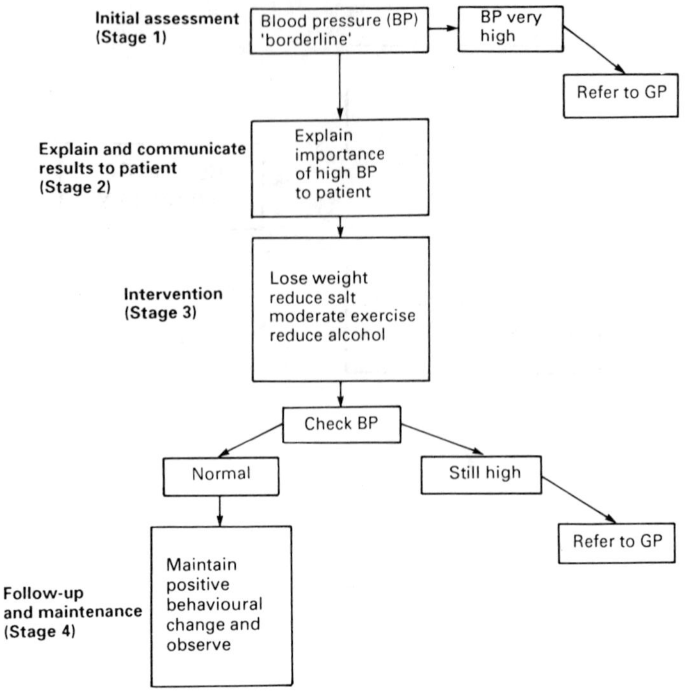

Fig. 9.2 *Blood pressure (BP) – screening and prevention*

Legal considerations in health screening

Legal issues arise at the recruitment stage and during employment. We shall consider these separately.

Pre-employment screening

There is nothing in general law to prevent an employer screening individuals prior to employment. This should not, however, be done on a discriminatory basis since, if it is, there may be a contravention of the Race Relations Act 1976 or the Sex Discrimination Act 1975. If the screening process reveals evidence of disease, or the *potential* for ill health, the employer may for that reason legitimately refuse to employ the individual in question. Generally speaking, the employer should ensure that the nature and purpose of the tests being carried out are fully explained to the person concerned, especially in relation to blood tests, and the AIDS test in particular. Written consent must be obtained prior to any examination and it should be made clear that the results are to be disclosed to the potential employer.

The Access to Medical Reports Act of 1988 sets out all of these principles. The act also stipulates that before an employer requests a report on an employee or prospective employee, from a medical practitioner, he must first obtain the consent of the individual concerned. The individual is also entitled to see the report, and to amend it if necessary, *prior* to it being seen by the employer. The individual may ask the doctor to amend the report, but if the doctor refuses then the individual has a right to attach his own written amendments to the report before it is shown to the employer. Indeed, the medical practitioner cannot supply the report to the employer unless the individual has had access to it or unless a period of 21 days has elapsed without the doctor having received any communication from the individual concerning arrangements for such access.

One exception to this is that a doctor is *not* obliged to give an individual access to a report if, in his or her opinion, its contents are likely to cause serious harm to the physical or mental health of that individual.

Finally, it is important to bear in mind that consent can be withdrawn at any time; so even if an individual initially agrees to undergo a medical examination, he may subsequently withdraw his consent. It may seem that the act works against the employer here, but, in practice, if a job applicant does withhold medical evidence, the employer is entitled to draw the obvious conclusion.

Screening during employment

The situation is fundamentally different once an individual

has been taken into employment. In general terms, an employer is only entitled to screen an employee if there is a prior contractual arrangement or where specific consent is given. If an employee refuses to undergo a screening or medical examination in contravention of a contractual arrangement, then this is likely to constitute a disciplinary offence which could result in dismissal. Moreover, theoretically the employer may also have the right to sue for breach of contract.

If no such contractual arrangement exists, then the employee is entitled to refuse to undergo the screening examination. An exception to this would be the case of an employer who believed an individual to be suffering from a condition which was making him unable to carry out normal duties or putting himself or his colleagues at risk. In such a case, even in the absence of a contractual right, the employer may have an implied right – based on his duty to other employees – to require testing or screening.

Whether a screening examination is carried out under a contractual agreement or purely voluntarily, the rights and obligations contained within the Access to Medical Reports Act still apply.

Buying a health screening programme

Screening programmes differ to some extent in content, as we have seen; the mode of delivery, however, varies enormously. Some screening organisations provide their services primarily via hospital-based programmes, some through specialist free-standing medical centres, and still others provide mobile screening units or 'on-site' screening programmes. Whatever the method of delivery, they all have one thing in common: they all cost *money*, even if the screening is done by your own medical department.

It is essential to be clear about what sort of screening

programme you require and what its short- and long-term objectives are to be. Have the answers to these preliminary questions before contacting any possible suppliers:

1. What do you wish to achieve from a health screening programme? For example, do you merely wish to provide an attractive company perk, or do you want more from it than that.

2. Whom do you wish to provide the service for? Do you wish to provide it for, say, the top 10 per cent of the workforce or do you wish to make it available to everyone.

3. What sort of population is it? In other words, what is the age distribution, and the male/female division? This will obviously affect what sort of screening tests are needed.

4. What sort of budget do you have available to provide this service?

When you do approach suppliers, you should seek answers to the following questions:

1. What are the medical and scientific justifications for the various components of the screening programme which they have recommended?

2. Are they able to target high risk groups and provide appropriate intervention strategies and follow-up?

3. Do they have sufficient back-up facilities if any of the individuals undergoing the screening process should require immediate specialist referral?

4. Are the doctors and nurses carrying out the screening appropriately qualified and trained in this type of work?

5. How do they report the results to the individual and

what is the range and quality of their health education literature?

6. Can they provide longer-term reinforcement of the health education message at regular intervals, which is important in ensuring lasting behavioural modification?

7. What sort of quality assurance programme do they run to ensure the continuation and improvement of standards?

8. What methods do they have of evaluating the impact of the health screening programme on the workforce?

This is not meant to be a comprehensive list of questions, but they should help you to crystallise your own thoughts as to what sort of programme you require as well as helping you to evaluate the services being offered to you.

CHAPTER 10

PREVENTING CARDIOVASCULAR DISEASE

In Chapter 3 we saw that the two principal forms of cardiovascular disease, coronary heart disease (CHD) and stroke, accounted for almost half of all deaths in England and Wales in 1985. In this chapter we shall be considering what strategies are available for the prevention of these diseases, and the effectiveness of such strategies.

In the normal state the arteries throughout the body are lined by a thin layer of cells called the *intima*. In the newborn child the lining of the arteries is perfectly smooth and regular, allowing normal blood movement throughout the system. With advancing age, and in the presence of certain risk factors, the lining of the arteries becomes thickened and irregular, a process which is primarily due to the deposition of certain fats, including cholesterol. The medical term for these deposits is *atheroma*, and when this process occurs in the arteries supplying the heart it is known as coronary heart disease (See Figure 10.1.)

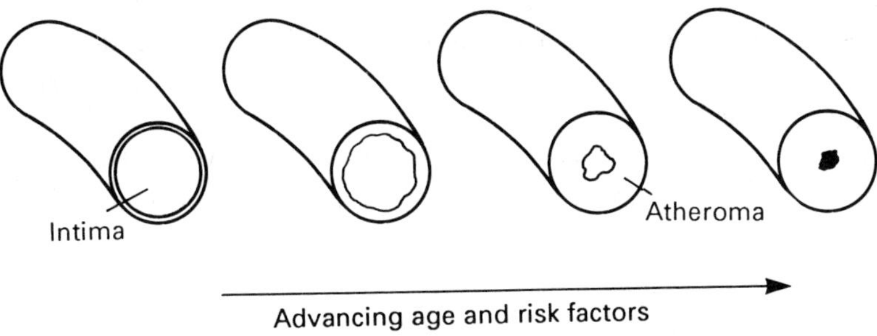

Fig. 10.1 *The development of coronary heart disease (CHD)*

What happens to people with heart disease?

When atheroma has developed to a point where 70 to 75 per cent of the normal diameter of the artery has been blocked, the individual may begin to experience symptoms. The most common of these symptoms is angina (pain across the chest and arms) and is due to insufficient blood being able to pass through the narrowed section of the artery. At rest there is usually no pain, but as soon as the individual tries to exert himself, for example, by walking up a steep hill, the pain starts. Sometimes a blood clot occurs in the narrowed section of the artery, completely blocking it, and that part of the heart supplied by the artery in question dies. Commonly known as a heart attack or coronary thrombosis, the medical term for this is 'myocardial infarction'. Figure 10.2 shows that the damage done to the heart will depend entirely upon which artery is blocked. Between 30 and 40 per cent of all major heart attacks prove fatal, with half of those deaths occurring within two hours of the onset of symptoms.

In many cases, heart disease causes sudden death, although no warnings or symptoms have been experienced by the patient.

What causes coronary heart disease?

Research has not identified a *single* cause for CHD. Indeed, there is unlikely to be a single factor which alone determines the development of the disease. There may well be a certain genetic predisposition involved since a family history undoubtedly increases risk, but it is unlikely that a single genetic marker could be used to distinguish between those who will develop CHD and those who will not.

Much of our understanding about the factors involved in causing heart disease comes from long-term observation of defined populations, after certain baseline measurements have been taken. Such studies have been carried out all over

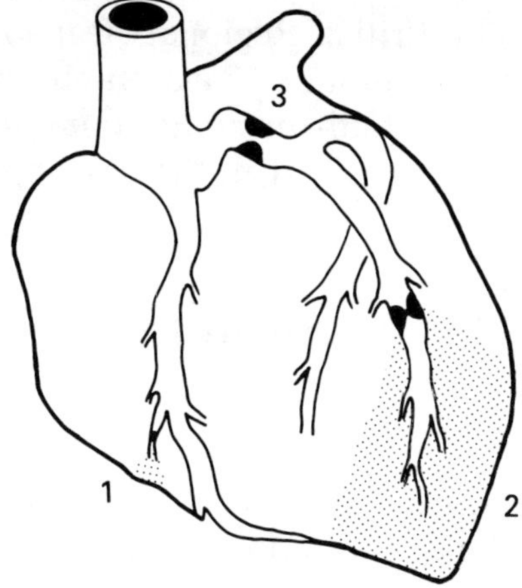

1. A relatively small artery is blocked and a small area of heart muscle has been damaged
2. Larger vessel involved and there is more extensive loss of heart muscle
3. A blockage at this level would threaten a very large part of the heart and may be lethal

Fig. 10.2 *The heart attack*

the world, but perhaps the most famous of them all is the Framingham Heart Study, carried out in the USA, which was described in Chapter 5. A similar study which was commenced in Britain about 10 years ago is called the British Regional Heart Study.

The British Regional Heart Study

The individuals recruited into this study were males aged 40 to 59 years, drawn at random from general practices in 24 towns from all the major regions of England, Wales and Scotland; 78 per cent of the men invited to take part appeared for examination between 1978 and 1980. They completed a questionnaire, and height, weight, blood pressure and respiratory function were recorded. An ECG was also taken and a variety of blood tests were done, including cholesterol concentration. Follow-up has continued and results are still being recorded. The British Regional Heart Study has produced a tremendous amount of extremely valuable information and has undoubtedly deepened our understanding of the causative factors involved in the development of heart disease.

What are the risk factors for heart disease?

Age and sex

The risk of heart disease increases with age and there is also a striking difference in male and female mortality from CHD. Men, on average, have a rate four to six times higher than women of the same age. This difference diminishes gradually with age, so that by the age of 80, death rates are virtually the same. (See Figure 10.3.) It seems likely that women are protected to some degree by the effect of specific female hormones; but, after the menopause, when the level of these is reduced, the risk of CHD increases significantly. It might be expected that replacing female hormones after the menopause (hormone replacement therapy or HRT) might provide some protection against CHD, and indeed some studies have suggested that this is the case.

The precise mechanism by which female hormones produce their protective effect is not known, but it is possible that they help to produce higher levels of HDL (see p 140) which appears to protect the arteries against the build-up of atheroma. The fact that women tend to have much higher levels of HDL than men tends to support this hypothesis.

A number of other risk factors have been identified as being associated with an increased tendency to develop CHD. (See Table 10.1.)

It is difficult to say which of the main risk factors is the most important. Complicated mathematical techniques, including the use of standardised regression coefficients, have been used in an attempt to identify the best predictor.

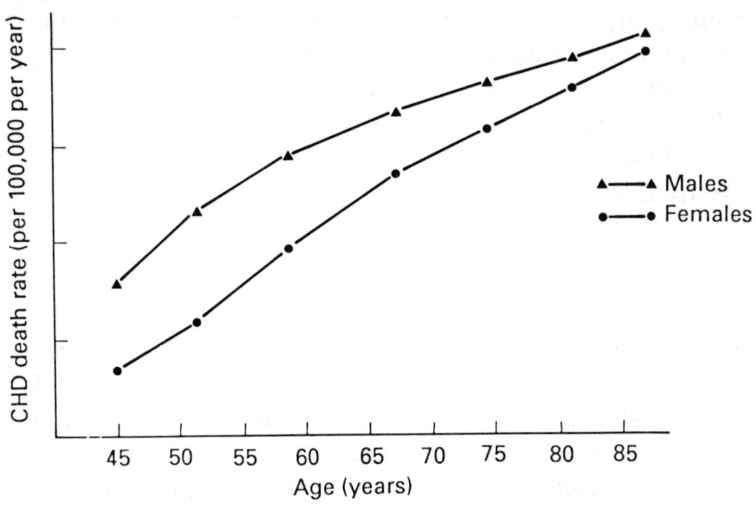

Fig. 10.3 *CHD death rates – males and females*

Table 10.1 *Risk factors for coronary heart disease (CHD)*

Risk factor	Effect on risk of CHD
Principal risk factors	
* *Cigarette smoking*	The more cigarettes smoked, the greater the risk
* *Blood pressure*	The higher the pressure, the greater the risk
* *Blood cholesterol*	The higher the cholesterol, the greater the risk
* *Physical inactivity*	The less exercise taken, the greater the risk
Secondary risk factors	
Family history	The longer parents live, the less risk for their children
* *Obesity*	Being overweight may increase the risk
Diabetes	People with diabetes have a higher risk
* *Stress*	Stress may increase risk in some people
Personality	Some personality types may be at greater risk than others
Hardness of water	The softer the tap water, the greater may be the risk

Source: adapted from 'Avoiding heart attacks' DHSS 1981
* Factors amenable to change

As a rough guide, evidence suggests that cigarette smoking is the most important factor in younger men, hypertension in older men, and hypertension, cigarette smoking and blood cholesterol concentration in middle-aged men.

All the principal risk factors and some of the lesser ones are amenable to change. There is absolutely no doubt that the individual can do a great deal to reduce the risk of CHD provided that he is given the knowledge and an appropriate plan of action. We shall now discuss each of these risk factors in a little more detail.

- *Cigarette smoking.* A number of studies have emphasised that the number of cigarettes smoked per day is a critical factor in CHD, with increasing risk associated with increasing consumption.
- *Blood pressure.* High blood pressure (hypertension) is associated with a significantly increased risk of CHD, with increasing levels being associated with increasing risk. It is extremely difficult to define 'normal' blood pressure, but the World Health Organisation gives it as a systolic pressure equal to or less than 140 mmHg, together with a diastolic value of 90 mmHg or less. (The higher of the two numbers is called the *systolic* pressure, which is the level reached when the heart is contracting most forcefully. The second and lower number is the *diastolic* pressure, and is the lowest value reached between contractions when the heart is at rest.) If 140/90 is the upper limit of normal, what should we regard as being abnormal or 'high' blood pressure? Again, it is difficult to be precise, but a sustained systolic pressure equal to or greater than 160 mmHg or a diastolic level equal to or greater than 100 mmHg would be regarded by most authorities as hypertensive. Blood pressure values *between* 140/90 and 160/100 are regarded as being 'borderline'.

  Evidence from the Framingham Study and the British Regional Heart Study shows at least a twofold increase in risk for the hypertensive group when compared with normal subjects. Studies performed in different populations have produced somewhat different levels of systolic or diastolic blood pressures at which risk becomes significantly increased. It follows that each community needs to ascertain its own levels of significance.

- *Blood cholesterol.* There is considerable evidence that the total cholesterol level is one of the most important single factors in determining the risk of CHD in individuals. Again, the level of risk rises progressively with increasing concentrations of blood cholesterol. Although the level of risk increases gradually to start with, at levels of 6.5 mmol/L and above, it increases dramatically. (See Figure 10.4.)

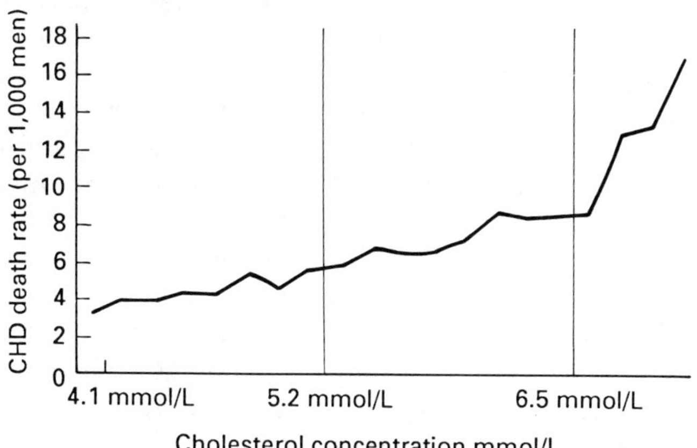

Fig. 10.4 *Cholesterol concentration and deaths from CHD*

A consensus view has now emerged as to what the optimum level of blood cholesterol concentration should be for the population as a whole. The British Hyperlipidemia Association has stated that a level of 5.2 mmol/L is what we should be aiming for.

- *Physical activity.* In Chapter 4 we discussed the importance of physical activity in some detail. Evidence strongly supports an inverse relationship between exercise and heart disease, that is, the less exercise taken the greater the risk and vice versa.

- *Family history.* It is generally accepted that a family history of cardiovascular disease, especially of CHD before the age of 50 in a first degree relative, constitutes an important risk factor for CHD; it may increase the risk two to four times.
- *Obesity.* The stereotype of a heart attack victim is a fat person, and there is a good deal of research data to support this picture. In the Framingham Study, overweight individuals appeared to have a twofold increased risk of CHD, compared with individuals of acceptable weight. The problem is that because obesity is so commonly associated with other risk factors, such as hypertension, diabetes and inactivity, it is not always easy to tease out exactly what contribution to risk obesity makes as an *independent* factor, that is, *apart* from its association with hypertension, diabetes, inactivity and so on. Whatever the mechanism, the message is clear: obesity matters!
- *Diabetes.* In the Framingham Study, the presence of diabetes was associated with a twofold increase in deaths from CHD, but the mechanism for this has yet to be discovered. Evidence suggests that diabetics are much more prone to the development of arterial disease in general, including the blood vessels supplying the kidneys and the eyes. Whether this is due to the diabetes itself or to some as yet undiscovered additional factor, remains to be seen. Interestingly enough, in the developing world, where CHD is uncommon, the presence of diabetes does not appear to produce an increased risk of CHD. This suggests that diabetes is not the fundamental causative factor, although it may aggravate the situation.
- *Stress.* We have discussed the question of stress

extensively in Chapters 8 and 13. Although the situation is complex, some individuals seem to be particularly prone to its effects and this may, in turn, increase the risk of CHD.

- *Personality.* There is some evidence to support the argument that certain personality types may be more prone to the development of CHD than others (See p220).
- *Drinking water.* There are some studies which suggest that those areas of the country with relatively 'soft' tap water have greater incidence of CHD. Whether these two factors have a causal relationship is not yet known.

Other risk factors

- *Sugar consumption.* It is commonly stated that sugar consumption is an important risk factor for CHD and other diseases. Indeed, Professor Yudkin has called it 'pure, white and deadly'. The evidence that sugar consumption is directly related to CHD is, however, thin.

  Nonetheless, as sugar is certainly associated with obesity, and obesity is in turn associated with hypertension, high blood fats, diabetes and inactivity, it would seem prudent to keep sugar consumption to a minimum.
- *Dietary fibre.* As we have seen in Chapter 5, there is some evidence that diets high in fibre are associated with lower blood cholesterol concentrations and a lower incidence of CHD. Whether this is due to a specific effect of fibre itself, or whether it is because diets high in fibre are almost invariably low in fat and thus produce lower cholesterol concentrations, is difficult to determine.
- *Coffee.* There is some evidence that coffee may have an indirect effect on CHD risk. In several studies, coffee drinking has been associated with an increase in total

cholesterol concentrations, and this would certainly tend to increase risk. There is also, of course, an association between coffee drinking and cigarette consumption. There is no association between tea drinking and elevations in blood cholesterol.

- *Alcohol.* There appears to be no direct relationship between alcohol consumption and risk of CHD. However, as mentioned previously, there is undoubtedly an association between excessive alcohol intake and disease of the heart muscle (cardiomyopathy), and there is a particularly strong relationship between excessive alcohol consumption and hypertension, which is a major risk factor for CHD.
- *HDL cholesterol.* We have seen (p 140) that HDL is a beneficial form of cholesterol which may help to remove fatty substances from the arterial wall. A number of studies have indicated that HDL cholesterol is associated inversely with the risk of CHD, that is, *low* levels of HDL are associated with an *increased* risk of CHD. Conversely, individuals with CHD consistently have lower concentrations of HDL than others. Factors tending to lower HDL include cigarette smoking and obesity, while physical activity is known to increase the level. Females generally have higher levels of HDL than males.
- *Salt.* There is no evidence that salt, in itself, contributes to risk for CHD. As with alcohol, however, increased consumption of salt is associated with increasing levels of blood pressure, so it may make an indirect contribution.
- *Oral contraception.* Oral contraceptives tend to increase blood pressure, increase blood fat concentrations and also to affect blood coagulation. So they are an indirect risk factor for heart attacks and strokes. This risk is significantly increased in the presence of other risk factors, particularly cigarette smoking and hypertension. The recent increase in CHD rates among younger women may be connected with oral contraception.

Multiple effects of risk factors

Risk factors combine in a synergistic way; that is, the combined effect of different risk factors exceeds the sum of their individual effects. So, for example, raised cholesterol concentration in conjunction with high blood pressure and cigarette smoking, increase the risk of CHD to more than eight times that for individuals without these factors. It follows that reduction of these risk factors from the levels currently prevailing in Britain, would be expected to yield benefit on a substantial scale.

There are two possible approaches to the prevention of CHD – the 'high risk' approach and the 'population' approach. In the former, individuals are screened in order to identify high risk due to the presence of known coronary risk factors. On the basis of this highly specific information, the individual can take appropriate action in order to reduce the risk of CHD to a minimum. The population approach seeks to reduce the incidence of CHD by lowering the mean level of risk factors in the population *as a whole.*

The high risk strategy for prevention

In this strategy, individuals participate in a screening programme specifically designed to identify known coronary risk factors. The results are analysed and interpreted, then individuals are given specific information and advice about their risk factors and how to reduce them. Then there is a follow-up and maintenance programme. We now consider each of these stages.

Baseline screening

This baseline screening does not require the involvement of a physician and can be carried out by an appropriately trained nurse or other paramedic. At the beginning is a brief questionnaire relating to the following factors: age, sex, cigarette consumption, previous history of heart disease or stroke, family history of heart disease, diabetes and exercise habits.

The following measurements should then be taken: height, weight, blood pressure and Body Mass Index (See p157). At this point it is useful to do a simple dipstick urine test to pick up evidence of kidney disease; it will also test positive for glucose (sugar) in the presence of diabetes.

Finally, a blood specimen should be taken for the measurement of blood cholesterol concentration. Contrary to popular belief, this does *not* have to be a fasting specimen. Under normal circumstances blood specimens are transported to a pathology laboratory. However, 'dry chemistry' machines now exist which can carry out blood cholesterol, blood glucose and other tests, on the spot. This has the clear advantage of being able to give the individual specific information about blood cholesterol concentrations immediately, without having to wait for a laboratory report. The difficulty is that these machines rely heavily upon the expertise of the user and are potentially open to considerable degrees of error. They also tend to lose accuracy at higher levels of blood cholesterol concentration, particularly above 7 or 8 mmol/L. The present advice is therefore to use one of the many excellent laboratories which have the necessary technology and expertise.

This initial baseline screening test should only take 10 to 15 minutes to complete.

Interpretation and risk stratification

As soon as all the information is available it can be interpreted and individuals can be classified according to level of risk (See p245). This is normally done by attaching a specific score to each risk factor and then adding the numbers up, although there are many other ways of measuring risk, some highly complex and requiring the use of computers. Giving individuals scores and overall risk factor assessments is an excellent way of motivating them. But this does not have to be on the basis of very sophisticated calculations. Table 10.2 shows a scoring system which is as good as any, and has the advantage of being quick and easy to use. You will see that the last section contains three brief questions on stress-related behaviour, which should be put to the individual at the time of the baseline screening. Individuals sometimes find it difficult to say which of these alternatives best describes their behaviour, which tends to vary from day to day. The important point is to choose the option which *on average* is the closest approximation.

It is extremely important that the results of the baseline screening should be communicated rapidly and clearly to each individual, so that full advantage can be taken of the intervention strategies.

Intervention and reducing the risk of CHD

Intervention strategies are designed to reduce the risk of CHD by encouraging individuals to adopt healthier lifestyle behaviour patterns. Giving them specific information about their own risk factor status, undoubtedly gives greater motivation and incentive to make such changes. Intervention programmes should be supported by good quality health education literature which can be produced in the form of a booklet or as an information pack containing separate sheets on each individual topic. The company can either produce its own literature or obtain literature already available from the Health Education Authority or the British Heart Foundation, for instance. This information should contain some 'interactive' element, that is, places in which an individual can record progress with regard to weight loss, exercise habits and so on. People like numbers, and they

Table 10.2 *Cardiovascular risk factor analysis*

			Score
1. Age	Age 56 or over	1	
	Age 55 or under	0	______
2. Sex	Male	1	
	Female	0	______
3. Family history	If you have:		
	Blood relatives who have had a heart attack or stroke **before age 60**	12	
	Blood relatives with a known history of heart disease **at or before age 60** but no heart attacks or stroke	10	
	Blood relatives who have had a heart attack or stroke **after age 60**	6	
	No blood relatives who have had a heart attack or stroke	0	______
4. Personal	**50 or under:** If you had either a heart attack, a stroke, heart or blood vessel surgery	20	
	51 or over: If you had any of the above	10	
	None of the above	0	______
5. Diabetes	Diabetes **before age 40** and now on insulin	10	
	Diabetes **at or after age 40** and now on insulin or pills	5	
	Diabetes controlled by diet **or** diabetes **after age 55**	3	
	No diabetes	0	______
6. Smoking	Two packs per day	10	
	Between one and two packs per day or quit smoking **less than a year ago**	6	
	If you smoke 6 or more cigars a day or inhale a pipe regularly	3	
	Less than one pack per day or quit smoking more than one year ago	3	
	Never smoked	0	______
7. Cholesterol	Cholesterol level – 6.5 mmol/L or greater	10	
	Cholesterol level – 5.2–6.4 mmol/L	5	
	Cholesterol level – less than 5.2 mmol/L	0	______
8. High blood pressure	If **either** number is:		
	160 over 100 (160/100) or higher	10	
	140 over 90 (140/90) but less than 160 over 100 (160/100)	5	
	If **both** numbers are **less than** 140 over 90 (140/90)	0	______
9. Body Mass Index (BMI)	The Body Mass Index is given by: $\frac{\text{weight in Kg}}{(\text{Height in metres})^2}$		
	BMI 30+	4	
	BMI 26–29	2	
	BMI 20–25	0	______
10. Exercise	Do you engage in any aerobic exercise (brisk walking, jogging, bicycling, racquetball, (swimming etc) for **more than 15 minutes:**		
	Less than once a week	4	
	1 to 2 times a week	2	
	3 or more times a week	0	______
11. Stress	Are you:		
	Frustrated when waiting in line, **often** in a hurry to complete work or keep appointments, **easily** angered irritable	4	
	Impatient when waiting, **occasionally** hurried, or occasionally moody	2	
	Comfortable when waiting, **seldom** rushed and easy going	0	______
	TOTAL POINTS		______

This is an adaptation of a scale developed by the Arizona Heart Insitute

Score results
Tablulate your points. Compare them with the charts below.

Please note! A high score does not mean you **will** develop heart disease. It is merely a guide to make you aware of a **potential risk**. Since no two people are alike, an exact prediction is impossible without further individualised testing.

Score	
High Risk	40 and above
Medium Risk	20–39
Low Risk	19 and below

enjoy having to record numbers on a day-by-day or week-by-week basis as evidence of their progress.

We now look briefly at each of the risk factors that are amenable to change. (See Part 2 for more detailed discussion of each factor.)

Cigarette smoking

Needless to say, all individuals should be encouraged to give up cigarette smoking, and it should be made clear that it is not only a major risk factor for cancer, but also for heart disease and stroke. It is always worth while giving up smoking, at any age; there is no truth in the argument that says – 'I have been smoking for so long now, it probably won't make any difference if I give up'. It is also a tremendous advantage to have a workplace smoking policy (see Chapter 6); it gives great encouragement to those smokers – and they are the majority – who *do* wish to give up. There is a wealth of educational information available from the Health Education Authority and other organisations, and full use should be made of it.

Blood pressure

The management of blood pressure can be exceedingly complex and, as we have seen, there is by no means universal agreement as to what levels of blood pressure require treatment. Table 10.3 sets out the three levels of blood pressure discussed earlier in this chapter, which we can use as guidelines. Below are the appropriate strategies for each group:

- Blood pressure 140/90 mmHg or less – *normal*: no specific intervention is required. Individuals should obviously be given appropriate health education concerning risk factors in general.
- Blood pressure greater than 140/90 but less than 160/100 mmHg – *borderline*. The individual should be encouraged to:

  – lose weight (exercise plus a low fat diet)
  – reduce salt intake
  – reduce alcohol consumption.

  Blood pressure should continue to be checked at regular intervals, and if there is no improvement, say, over a three-month period, even despite reduced alcohol and salt intake and weight loss, the individual should be referred to his general practitioner for further monitoring and management.
- Blood pressure 160/100 or greater – *hypertension*. Individuals should be referred to their general practitioner for further advice and management. However, they must also be encouraged to reduce weight, reduce salt intake and reduce alcohol consumption.

Table 10.3 *Blood pressure (BP) classification*

	Blood pressure (BP) classification
Normal	A blood pressure of 140/90 mmHg or lower
Borderline	A systolic pressure greater than 140 but less than 160 mmHg A diastolic pressure greater than 90 but less than 100 mmHg
Hypertension	A systolic pressure of 160 mmHg or greater A diastolic pressure of 100 mmHg or greater

Blood cholesterol concentration

The optimum level for the population as a whole is said to be 5.2 mmol/L. Similar levels have been adopted by many other countries, including the United States of America. This is a useful figure despite the fact that, strictly speaking, one cannot pinpoint a level of cholesterol which is best for the whole population. We know that the value of intervention diminishes with increasing age, and also that for any given cholesterol concentration the risk for a man is greater than for a woman.

Intervention strategies for the three levels of cholesterol concentration outlined previously are as follows:

- Blood cholesterol concentration 5.2mmol/L or less: No specific intervention is required, but individuals should be given general advice regarding healthy eating as a matter of course.
- Blood cholesterol concentration greater than 5.2 but less than 6.5mmol/L: These individuals should receive general advice on healthy eating patterns, and particularly recommended to reduce total fat consumption. It is not necessary to advise specifically on avoiding high cholesterol foods. Advice on other coronary risk factors should also be given.
- Blood cholesterol 6.5mmol/L or greater: These individuals should be given a specific cholesterol lowering diet. Advice regarding other coronary risk factors should also be given.

There is an important, although uncommon, exception to this general rule. Laboratory investigations normally measure *total* cholesterol concentration which includes the beneficial HDL cholesterol. The British Hyperlipidaemia Association has recommended that individuals with blood cholesterols above 6.5 mmol/L should receive specific intervention *except* when the raised

> total cholesterol concentration is primarily due to an increase in HDL (above 2 mmol/L). In practice, levels above 2 mmol/L for HDL are not common, especially in males. This means that recommending intervention for blood cholesterols at 6.5 and above is an appropriate method of management. If, after a three- to six-month period, when a full lipid profile (including HDL) can be carried out, the HDL is found to be above 2mmol/L, then the individual does not need to adopt so strict a dietary regime. If, on the other hand, the HDL is below 2 mmol/L then he should stick to the cholesterol lowering diet.
>
> If after a period of three to six months, the blood cholesterol has not responded adequately, then consideration needs to be given to the question of drug treatment. Some doctors start treating at a level of 8 mmol/L or above, while others start treatment at around 10 mmol/L or above. In any event, drug treatment should always be accompanied by changes in diet.
>
> Whatever reduction method is used, the target cholesterol concentration is 5.2 mmol/L. This is particularly important for younger patients, for those with genetic forms of raised blood cholesterol (familial hypercholesterolaemia), for those with a family history of coronary disease, or in the presence of other risk factors such as hypertension and diabetes.

In the United States the recommendation is that total cholesterol should be measured in all adults, 20 years of age and over, at least once every five years. In this country the British Hyperlipidaemia Association has recommended that all adults should have their blood lipid concentrations measured at least once, preferably before the age of 30.

It is important to emphasise that individuals with blood cholesterol concentrations between 5.2 and 6.5 *are* at increased risk and should try to reduce levels to the optimum of 5.2 mmol/L. In the British Regional Heart Study, 20 per cent of acute CHD events (fatal and non-fatal) occurred in the lower 40 per cent of the cholesterol distribution (less than 6 mmol/L).

The British population has a high average level of total cholesterol (about 6.3 mmol/L) and as a risk factor it is widely distributed. But the fact is that most people who have high cholesterol concentrations do not develop heart disease, although most patients with CHD do have high levels of total cholesterol. It follows that other factors must determine the development of CHD in individuals who are already at risk due to elevated cholesterol concentrations. These additional factors are principally smoking and hypertension, but other risk factors for CHD are also important. On the basis of current evidence, it seems that total cholesterol concentration is the major factor which identifies a high risk group, while other factors such as smoking and hypertension create a subgroup who go on to develop CHD. In managing elevated blood cholesterol levels, it is therefore of the utmost importance to pay close attention to other important coronary risk factors.

Obesity

Individuals with a Body Mass Index (see p 157) of 30 and above (classified as obese) have a significantly increased risk of CHD, and even those with more modest degrees of overweight are also at increased risk. Obesity is important because it is often associated with many other risk factors, including hypertension, diabetes, high blood cholesterol levels and, of course, inactivity. Weight loss can have beneficial effects on blood pressure, blood fat concentrations and it allows better control of diabetes.

Many people are in the business of persuading us that there are ‘fast’ and ‘effortless’ solutions to weight problems. The fact remains that if you take in more calories than you expend you will put on weight. If you expend more calories than you take in, you will lose weight. If you take in the same number as you expend, your weight will remain the same. It follows that in order to lose weight one must reduce the *intake* of calories, increase the *output* of calories, or (preferably) do both. The best way, in brief, to lose weight is to combine a low fat, high fibre diet with a regular programme of moderate exercise. Weight loss of two to three pounds per week is likely to be more beneficial than trying to lose a lot of weight over a short period of time. (Crash diets do not remove fat weight, they remove glycogen and water from muscle, which gives an illusion of rapid weight loss.)

It is important to weigh individuals regularly, to give them feedback and encouragement, and this can easily be done in the workplace without any medical supervision. Small groups of people with weight problems can meet together on a regular basis, and this has been shown to be a much more effective way of dealing with overweight and obesity than individuals trying to cope alone. Peer pressure, and peer encouragement are extremely important factors in determining success for most people.

Stress

Stress is discussed in some detail in Chapters 8 and 13. Although stress may not act alone as a primary cause of CHD, there is no doubt whatever that prolonged stress substantially increases the risk of CHD. Prolonged and unremitting stress in an already high risk individual can prove lethal. It follows that techniques designed to deal with stress at an organisational and an individual level are an essential component of an overall risk management strategy.

Diabetes

Diabetes increases the risk of CHD which makes it all the more important for the diabetic to control any other coronary risk factors such as smoking and hypertension. In older patients who have the maturity onset form of the disease, diet alone may be sufficient to keep the disease under control. This is always the best option since all drugs have side-effects of one form of another.

Lack of physical exercise

In Chapter 4 we discussed the importance of regular physical exercise to the prevention of CHD. Individuals should be encouraged to participate in physical activity to improve the performance of the cardiovascular system. The form of exercise chosen must involve major muscle groups, and be regular and rhythmical. Brisk walking, running, cycling, swimming and tennis are all examples of this form of exercise, and current evidence suggests that three or four sessions of 20 to 30 minutes each week are sufficient to promote cardiovascular fitness and to confer a significant measure of protection from CHD, via a variety of important physical and psychological effects. As in all other cases, close attention should be paid to any other coronary risk factors which may be present.

Additional screening techniques

Total cholesterol/HDL ratio

We know that the higher the total cholesterol concentration, the greater the risk of CHD. We also know that the higher the level of HDL concentration, the *lower* the risk of CHD. Some authorities believe that it is better to measure both parameters since HDL is as good a predictor of CHD events as is total cholesterol. The relationship between these two risk factors can be expressed as follows:

$$\frac{\text{Total cholesterol}}{\text{HDL}} = \text{CHD Risk Ratio.}$$

This calculation takes into account the protective effect of HDL as well as the risk effect of total cholesterol, and an average risk ratio is about 5. (See Table 10.4.) The higher the risk ratio, the greater the risk, and even small increases or decreases in HDL can affect it substantially. For example, if the total cholesterol for an individual were 8 mmol/L per litre (which is high) but the HDL fraction was 2, then the risk ratio is given by 8 ÷ by 2 = 4. In other words, even though the cholesterol concentration is significantly elevated, the actual risk of CHD is *below* the average level of 5. In this situation, although it would be useful for the individual to reduce total cholesterol it would perhaps be not as important as if the HDL were only 1. In this case the risk ratio (8 ÷ 1 = 8) is clearly *above* the average level, and the individual needs to reduce total cholesterol concentration by sticking closely to a specific diet. Reducing total cholesterol by only 2 mmol/L (from 8 to 6), as a result of dietary intervention, would give a CHD risk ratio of 6 ÷ 1 = 6, which is only just above the average of 5.

The benefit of taking HDL into account when providing individuals with advice about diet, is that the advice can be much more specific. Measuring total cholesterol alone is a perfectly good initial screening test on the basis of which reasonable advice to lower cholesterol concentration can be given. Measuring the HDL at the same time shows the degree of risk more accurately and advice to lower the total cholesterol concentration can be reinforced in those individuals whose CHD risk ratio is greater than average.

The problem is that the measurement of HDL is a much more expensive investigation. It would therefore seem reasonable only to measure HDL concentrations in those who are initially found to have high total cholesterol concentrations.

The best prevention strategy as far as cholesterol is concerned can be summarised as follows:

Table 10.4 *Total cholesterol/HDL ratios compared with the level of risk*

	Risk	Total Cholesterol HDL Cholesterol
Men		
	½ average	3.4
	average	5.0
	2 × average	9.5
	3 × average	23.4
Women	½ average	3.3
	average	4.5
	2 × average	7.0
	3 × average	11.0

1. Measure total cholesterol concentrations in everyone.

2. Identify those who are at high risk, and advise a specific cholesterol lowering diet and management of other coronary risk factors.

3. After three to six months, test high risk individuals' total cholesterol concentration again to determine the response to the diet. At the same time measure HDL concentration. This will allow calculation of the CHD risk ratio.

4. Individuals whose CHD risk ratios are below average do not need to remain on a specific cholesterol lowering diet. They should be given general dietary advice and do not need to be followed up.

Those individuals whose CHD risk ratios are above average should continue on the diet and should also be given advice concerning other coronary risk factors. Further measurements of total cholesterol and HDL should be carried out periodically to ensure adequate response. If it is not adequate, individuals should be referred to their GP for further management and advice.

Exercise stress testing

The technique of exercise electrocardiography has been around for many years and is used mainly in the investigation of patients with cardiac disease. The principle of the test is that when the diseased heart is under load, that is, during exercise, electrical changes may show up on the ECG which were not present in the resting state. A *resting* ECG is of very limited value in the identification of CHD since it may be entirely normal even in patients with advanced disease. The exercise ECG, on the other hand, yields a great deal more information and it may well show evidence of advanced disease even in the absence of symptoms. This is an important point because there is still a commonly held view (even among some doctors) that 'no pain means no disease'. But it is quite possible to have serious heart disease without any symptoms at all, which explains why many individuals die suddenly having had no advanced warning of CHD.

In recent years a number of commercial organisations have included exercise stress testing as part of a comprehensive screening programme. Some critics have argued that this sort of testing in an apparently healthy population, does more harm than good because it produces a large number of false positives (patients who are told that they do have disease when they in fact do not). But as with other techniques the crucial factor is the competence of the operator. The exercise electrocardiograph should never be interpreted in isolation, but must always take into account the patient's clinical status and the presence of other known coronary risk factors. It should always be carried out by an experienced operator and full resuscitation facilities should always be available.

Even an exercise ECG cannot exclude underlying CHD with absolute certainly. All one can say is that if a maximal exercise stress test is normal, it is powerful evidence *against* the presence of underlying CHD. However, its predictive power is much enhanced if it is used primarily in high risk groups. The more coronary risk factors are present, the greater the probability of underlying CHD, and therefore the greater the probability that the test will yield positive results. It is important to identify individuals who already have advanced coronary disease, even if they do not yet have symptoms, because newer techniques, such as coronary angioplasty, facilitate early intervention and offer better prospects for long-term survival. In expert hands, the exercise ECG in an immensely valuable screening instrument.

In addition to providing valuable information about heart function, the test can also yield important information about the individual's aerobic capacity (fitness). The fitness level of the subject can be determined approximately by measuring performance on a treadmill. But nowadays there are much more sophisticated and accurate means of assessing fitness. One technique, which is increasingly used, is to measure exercise Repiratory Exchange Ratios, for example Maximum Oxygen Consumption. These measures give precise indications of an individual's fitness level and also take into account factors such as body weight. The more advanced systems allow measurement of these parameters on a breath-by-breath basis (each breath is analysed individually).

The important point is that, as well as helping to diagnose underlying CHD, exercise stress testing is an extremely useful means of encouraging individuals to improve their aerobic capacity by regular moderate exercise. Since we are trying to encourage a relatively sedentary population to become more active, this effect of the exercise testing process should not be undervalued.

The population approach to prevention

Instead of attempting to identify a high risk group, the

population approach seeks to reduce the incidence of CHD by lowering the average level of risk factors in the population *as a whole*. Advocates of this approach argue that coronary risk factors are so widely distributed in the population, that it is better to direct resources towards reducing the risk for everyone, rather than focusing upon the high risk individuals. There is some evidence to support this view. In the UK Heart Disease Prevention Project, the top 15 per cent of the risk distribution, only yielded about one third of the total number of individuals who eventually developed CHD. The British Regional Heart Study produced evidence which was remarkably similar: the top 20 per cent of the risk distribution only yielded about one third of the total number of subjects who went on to develop heart disease. This implies that a large number of people at *moderate* risk may give rise to more cases of disease than the small number who are at high risk. A preventive strategy designed to raise the level of awareness of health issues in the whole population, without specifically seeking to screen out the high risk groups, may therefore be a better use of resources. From the Framingham data it has been calculated that a 10 mmHg lowering of the blood pressure levels across the population as a whole, would be reflected in a 30 per cent reduction in total deaths from heart disease. Another estimate is that a 10 per cent reduction of blood cholesterol would reduce CHD by as much as one third.

High risk approach v population approach

The two approaches to the prevention of CHD discussed here each have their advantages and disadvantages; neither one is 'better' than the other, and the most potent strategy will involve the best elements of *both*.

The major problem with the population approach is that it yields only a small measure of benefit to each individual. While it is true that persons who are only at moderate risk of

CHD will ultimately generate more cases than those who are classified as high risk (simply because there are more of them), the fact is that most of the moderate risk group, can expect to remain well. The great advantage of the high risk approach is that instead of being on the receiving end of rather anonymous initiatives designed for the population as a whole, the screening process allows the individual to be given highly specific information about his or her risk factor status. This has been shown to be an extremely powerful and effective means of motivating those who are most at risk to take effective remedial action.

Nevertheless, the population approach is of crucial importance in our longer-term strategy of prevention, because it seeks to change fundamentally the attitudes and behaviour patterns of society as a whole. This is especially important for the younger members of our society, including the children in our schools. Looking to the future, they are the ones who are most at risk and it is of the utmost importance that we should reach them before unhealthy lifestyle practices and behaviour patterns have become fully established. Indeed, the main beneficiaries of the population approach will be those who are young – or not yet born – today.

The high risk approach will be of more immediate benefit to the present adult population. In the screening programme outlined above, it is not only those at high risk who receive counselling; those in moderate and low risk categories also receive information concerning effective risk factor management. It is clear that in combining the population approach and the high risk approach, the former underpins the latter. A report from the British Cardiac Society describes this twin approach as 'altering the mass characteristics of lifestyle which are the underlying causes of mass disease and identifying and helping individuals at special risk within the population'. These, it says, are essential elements of a prevention strategy.

Does prevention work?

A large number of studies have been carried out in an attempt to answer this question and they all point towards the same conclusion: that changes in personal behaviour or lifestyle (stopping smoking, taking exercise, reducing weight, dietary changes and so on) can lead to a significant reduction in the incidence of CHD. Long-term studies involving large numbers of people are extremely difficult to administer and hugely expensive. But their cumulative results provide sufficient information and data upon which to mount an overall strategy for CHD prevention.

Coronary rehabilitation

It would be unthinkable to end this chapter without some mention of those patients who have already suffered their first heart attack. There are a substantial number of such individuals for whom prevention (in this case tertiary prevention, that is prevention when the disease has already caused symptoms) is even more important. Although attitudes towards patients who have suffered heart attacks are now more liberal than they used to be, there is still misunderstanding about the disease. Many people tend to ostracise post-coronary patients and regard them with apprehension. In consequence, patients lose confidence and self-esteem; the social effects on the family can be catastrophic. The problem of heart disease should not be accepted; it should be attacked.

The future for the post-coronary patient depends primarily upon two factors: the amount of healthy heart muscle which remains; and the condition of the arteries supplying that muscle. The aim of coronary rehabilitation is to develop the function and efficiency of the remaining heart muscle as much as possible, and to do everything to reduce further development of disease in the arteries which supply that muscle. The former is achieved through regular controlled exercise and the latter by close attention to known coronary risk factors such as smoking, elevated cholesterol levels and hypertension. The intended aim of such a programme is to return the patient to the highest possible level of physical and psychological activity.

Organised programmes of cardiac rehabilitation are not widely available in this country; unlike the USA and some European countries. Why this should be so is not entirely clear except that it probably reflects the high degree of apathy which hampers most attempts at prevention. Most post-coronary patients, however, are highly motivated to change their lifestyles in an effort to reduce risk. Many can return to a normal and productive life, often in better physical and psychological condition than they were prior to their coronary attack.

Strokes

The essential underlying mechanism in coronary heart disease is the progressive narrowing of the coronary arteries by fatty deposits called *atheroma*. In cerebral vascular disease the same process of arterial degeneration and blockage occurs in the vessels supplying the brain, but while complete blockage of an artery supplying the heart causes a heart attack (myocardial infarction), in the brain the same event will lead to a cerebral infarction or stroke. Stroke is the term employed to describe the severe acute manifestations of cerebral vascular disease which result from interruptions to the normal blood flow in the brain. The clinical manifestations of stroke are wide ranging but normally involve some degree of paralysis to one or other side of the body and varying degress of disturbance in balance, sensation and speech.

About 70,000 people each year in England and Wales die as a result of a stroke and this is equivalent to about one quarter of all mortality from diseases of the circulatory system. In 1986 more than 70 per cent of fatalities attributed to strokes occurred in individuals aged 75 years and over. But this does not mean that stroke is an unimportant cause of premature death and disability in younger age groups. Of the 27,000 or so males dying from stroke in 1986, 10 per cent (2,700) occurred in the age range 55 to 64. In females in the same year, of the 43,000 or so deaths, about 5 per cent (2,150) occurred in the age range 55 to 64. In other words, in 1986, more than 4,800 deaths from stroke occurred in individuals in the 55 to 64 year age group.

As with CHD, there is no single specific cause of stroke, but a number of risk factors have been identified which are associated with a significantly increased risk of developing stroke in the future. These factors are as follows:

- *Hypertension.* Hypertension is undoubtedly the most important risk factor for the development of stroke. As with heart disease, as blood pressure increases, so the risk increases, and as we know from our previous discussion, hypertension is an extremely prevalent condition – the world health organisation estimates that as many as 10 to 20 per cent of the population may be hypertensive.
- *Cigarette smoking.* As with heart disease, cigarette smoking is a significant risk for the development of stroke and this should come as no surprise. The most recent evidence from the Framingham Study confirms this adverse effect and also demonstrates a dose – response relationship, that is the risk of stroke increases as the number of cigarettes smoked increases.
- *Heart disease.* The presence of CHD may double the risk of stroke, which may be partly explained by the tendency of diseased hearts to develop abnormal

rhythms which may result in blood clots affecting the brain (emboli).

- *Diabetes.* As in CHD, diabetic patients are at a higher risk for the development of stroke. Evidence suggests that diabetes as a risk factor is more significant in women than in men, although the reason for this is not entirely clear.
- *The contraceptive pill.* The use of oral contraceptives appears to be a risk factor for strokes as it is for heart disease. The mechanism for this is presumably related to the fact that the contraceptive pill tends to increase blood pressure and also increases the tendency of the blood to clot. A particularly dangerous combination is the use of the contraceptive pill and cigarette smoking.
- *Alcohol.* Because of the association between excess alcohol consumption and hypertension, it should come as no surprise that alcohol is a significant risk factor for the development of stroke. Recent studies suggests that heavy drinking males have about four times the risk of stroke compared with non-drinkers. Moreover, this increased risk persists even when allowances are made for cigarette smoking, hypertension and other risk factors, and the evidence therefore strongly suggests that a high blood alcohol intake is an *independent* risk factor for stroke that can and should be prevented.

Preventing stroke

Since the most important risk factor for the development of stroke is hypertension, it follows that the most effective means of prevention is to identify those who are hypertensive and provide appropriate intervention, which we discussed in relation to CHD.

Worksite blood pressure measurement is a simple, inexpensive investigation and should be made available to all employees. Those with a BP of 140/90 or less need no specific intervention, while those with levels above this need specific advice on reducing salt consumption, reducing weight and cutting back on their alcohol intake. Those with levels above 160/100 should be referred to their own general practitioner for additional advice.

All employees should receive advice regarding cigarette smoking, the need for regular exercise and the importance of modifying their intake of saturated fat.

CHAPTER 11

TACKLING ALCOHOL AND DRUG ABUSE IN THE WORKPLACE

ALCOHOL ABUSE

A strategy designed to reduce the potentially damaging effects of alcohol on the workforce, can be considered under five headings:

- Educating the workforce
- Identifying the problem drinker
- Identifying the causes
- Management of the problem drinker
- Establishing an overall alcohol policy.

Educating the workforce about drink

There appears to be a surprising amount of ignorance about the nature of alcohol and its potentially damaging effects, so an education programme is an extremely valuable frontline component of any preventive strategy. Possible methods to use are leaflets/booklets, posters, videos, discussion groups/ seminars and health education exhibitions. Whichever method or combination of methods is chosen, it must get across information on the following:

1. The alcohol content of different drinks and the concept of alcohol 'units' (see p 194–196)
2. The damaging physical and psychological effects of prolonged alcohol abuse
3. The social aspects of alcohol abuse, including drinking and driving

4. The effects of alcohol on performance at work and on the risk of accidents and injury
5. Safe drinking limits for men and women and guidance on how to use alcohol sensibly
6. The company policy on alcohol and, in particular, a contact point where employees can obtain help and advice if they feel that they have a problem. In this context, it is imperative to stress complete confidentiality and the company's understanding and supportive attitude.

Training and educating managers and supervisors

Specific training is required for senior management and supervisors, since they will be the ones responsible for implementing an overall alcohol policy. The training programme needs to address in more detail the following issues:

1. The nature of alcohol abuse, its possible causes and effects
2. The implications of alcohol abuse in the workplace in terms of efficiency, productivity and, in particular, safety
3. The recognition of the problem drinker
4. The company alcohol policy and overall procedures for managing alcohol abuse, including counselling and more specialised help from outside agencies.

Identifying the problem drinker

The two principal approaches to the identification of the problem drinker are set out in Figure 11.1 and discussed below.

General appearance, behaviour and work performance

Supervisors and managers should be able to recognise

certain signs which suggest that someone might be drinking too much. None of these changes in appearance or behaviour *necessarily* indicate an alcohol problem; there are other possible causes. Nevertheless, signs such as the following should alert managers or supervisors to the possibility that an alcohol problem exists:

- Repeated lateness and absenteeism
- Deteriorating personal appearance, smell of alcohol and unusual behaviour
- Inefficiency and reduced productivity
- Conflict and deteriorating relations with working colleagues
- Poor concentration, memory and judgement
- Accidents

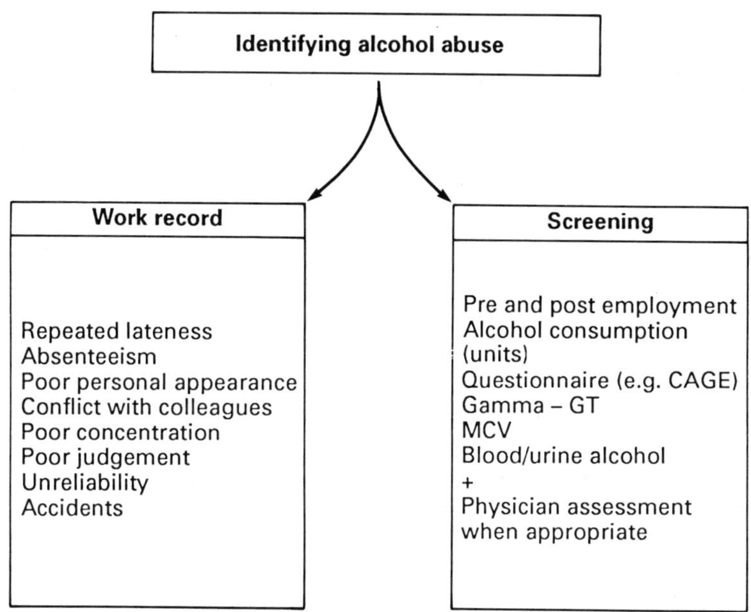

Fig. 11.1 *The two basic approaches to identifying alcohol abuse in the workplace*

It is extremely important to document any evidence clearly and precisely. Individuals with alcohol problems have an extraordinary capacity to deceive themselves and, more often than not, attempt to deny or to distort the truth. At some point the individual has to be confronted with the evidence and, although this must be done in a non-threatening way, the evidence must be irrefutable.

Screening for alcohol abuse

Screening for alcohol abuse may be carried out either pre-employment or post-employment. (For the legal aspects of health screening, see p248.) Pre-employment screening for alcohol abuse may be carried out as part of the normal pre-employment medical examination. Post-employment screening may be carried out in two situations: as part of a routine screening programme made available to all or part of the workforce; and where there is already strong suspicion that an individual is abusing alcohol.

There are a number of screening procedures available but most use a combination of a clinical history and examination, a questionnaire, and a variety of blood tests.

Clinical history and examination

A physician carrying out the screening examination should look at the individual's previous and current alcohol consumption, the work record, including absenteeism and accidents, and any family history of alcohol abuse. The doctor should be experienced in the identification of the wide variety of physical and psychological manifestations of alcohol abuse, such as hypertension, tremor, muscle wasting and liver enlargement.

Questionnaire

A variety of questionnaires have been used to try and identify the problem drinker. Such questionnaires commonly employ a variety of different social and psychological parameters to separate harmful from problem-free drinking. The more widely used brief screening tests include the MAST, the STAQ and the CAGE questionnaires.

The CAGE questionnaire consists of only four questions, from which the acronym is derived. They are as follows (1) Have you ever felt you ought to *Cut* down on your drinking? (2) Have people *Annoyed* you by criticising your drinking? (3) Have you ever felt bad or *Guilty* about your drinking? (4) Have you ever had a drink first thing in the morning to steady your nerves, or get rid of a hangover (*Eye*-opener)? Two or more positive replies are said to identify the problem drinker. In the original study of 366 American psychiatric patients, 81 per cent of known alcoholics answered positively to two or more questions, compared with 11 per cent non-alcoholics.

The MAST (Michigan Alcoholism Screening Test) is a 25-item interview comprising questions related to personal opinions on drinking, opinions of family and friends, problems arising from drinking and some symptoms of alcohol dependence. A shortened version of the MAST is available which comprises the ten most discriminating items from the original test. The questions asked are direct and unequivocal in their focus on alcohol. They require the respondent to admit that he has a drinking problem.

The STAQ (Spare Time Activities Questionnaire) is longer than the instruments described above, requiring seven to eight minutes to administer, but its ability to identify alcohol abuse seems to be no better than the briefer questionnaires.

The CAGE questionnaire and others like it are useful in detecting heavy drinkers and problem drinkers, but are less reliable in detecting dependent drinkers (alcoholics). This is because dependent drinkers often lie in answer to questions relating to their alcohol consumption. On the other hand, many excessive and problem drinkers are surprised that they should be considered as such, and the CAGE questionnaire is often an excellent means of driving home the point that they really are at risk.

Blood tests and other tests

Blood tests have the advantage of reliability and objectivity, although some are very much influenced by the time interval between an individual's last drink and the time of taking the sample.

One of the most commonly used measures is that of Gamma-Glutamyl transpeptidase (Gamma-GT). This appears to be an extremely sensitive test for identifying early liver disease. Some studies have reported an elevation of Gamma-GT in 60 to 80 per cent of alcoholics, although it is probably a less sensitive indicator in a general health screening programme. The presence of an elevated Gamma-GT does not confirm the diagnosis of alcohol abuse, it merely indicates liver damage. The vast majority of cases do turn out be due to alcohol abuse, but one should not ignore other potential causes. The laboratory which you are using to carry out these investigations will give you detailed advice and will also supply you with appropriate reference or 'normal' ranges for Gamma-GT. (Reference values vary between laboratories, depending upon the specific method used and a variety of other factors.)

Elevated levels of Gamma-GT begin to return to normal after about 48 hours of abstension from alcohol, and therefore the time of blood sampling is critical. Some studies have suggested that women show a proportionately rising Gamma-GT in response to equivalent amounts of alcohol, suggesting that they may be more sensitive to alcohol abuse than men.

Another blood test commonly used is related to the size of

the red blood cells and is called the mean corpuscular volume (MCV). An enlargement of the red blood cells and thus a raised MCV is commonly found in excessive drinkers, and MCV is now a widely used indicator of harmful drinking. Current evidence suggests that the MCV is raised in 50 to 60 per cent of problem drinkers.

Anyone who drinks regularly and heavily develops tolerance to the effects of alcohol, and individuals who appear sober and well co-ordinated at the time of interview, may nevertheless have significant levels of blood alcohol. Thus the value of blood or urine alcohol measurements should not be overlooked as a possible guide to alcohol abuse. Nowadays the use of a quantitative breathalyser is simple and rapid and is likely to come into more general use as a screening instrument.

Screening a large number of employees using a physician-based programme is expensive and time consuming. A more rational use of resources would be to use a nurse-based screening programme (under the supervision of a doctor). Those identified as at risk can then be referred to a doctor for more detailed clinical evaluation and assessment. (See Figure 11.2.) Obviously, any individual already under strong suspicion of alcohol abuse should be referred immediately for expert counselling and advice, and in most cases this will include a medical opinion.

Do screening and intervention make any difference?

It is appropriate to ask whether health education and simple screening measures implemented by a company actually make any difference to alcohol consumption and alcohol-related disease. There are a number of studies which strongly suggest that such activities make a significant difference and we shall discuss two of them in this section.

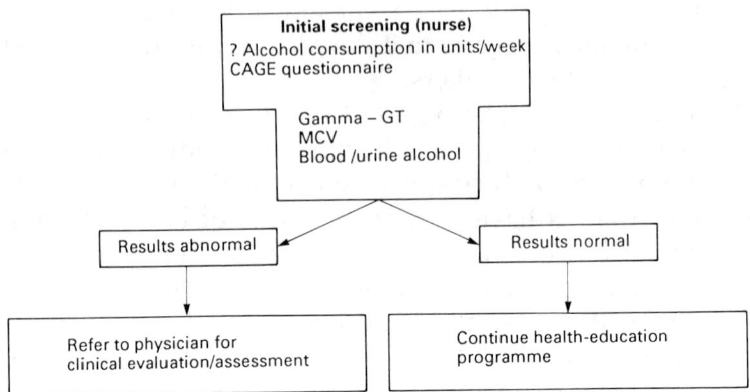

Fig. 11.2 *Basic screening programme to detect alcohol abuse*

Edinburgh royal infirmary project

A study was undertaken in Edinburgh to assess the effectiveness of brief intervention. A number of patients in a general hospital were identified as having a current alcohol problem. Identification was done by a nurse-based programme using a structured interview covering drinking habits, medical history and social background. The MCV and the Gamma-GT were recorded in each case.

The problem drinkers were then divided into two groups, an intervention group and a control group. No advice was given to the control group, although they all agreed to a follow-up interview one year later. The intervention group, on the other hand, received further counselling from the nurse, and this was supported by a booklet containing advice about ways to reduce alcohol consumption.

After one year both groups were interviewed by a nurse who did not know the design of the original study. Findings indicated that fewer of the intervention group had alcohol problems at the end of one year (35 per cent) compared with the control group (62 per cent). This evidence is clearly encouraging, and demonstrates that relatively simple intervention by an experienced nurse does have a positive effect on drinking problems during the course of the ensuing year.

The Malmo project, Sweden

The Swedish study involved 8000 middle-aged men in Malmo who underwent routine health screening. From this initial sample 585 showed evidence of heavy drinking on the basis of an elevated Gamma-GT. The 585 men were randomly divided into two groups, an intervention group and a control group. The control group was simply informed by letter that they had evidence of impaired liver function and they were advised to cut down on their alcohol consumption. They were invited to return for further blood tests after a two-year period.

The intervention group, on the other hand, was given a detailed physical examination and interviewed about their drinking history, symptoms of alcohol dependence and evidence of alcohol-related problems. This group was then offered appointments with the same doctor every third month and monthly contacts with the same nurse who originally carried out the Gamma-GT test. Individuals were given detailed advice about how to moderate their drinking habits and progress was monitored by regular feedback of Gamma-GT levels. Once the Gamma-GT had returned to an acceptable level, the frequency of clinical contact was reduced.

The progress of both groups was evaluated at two and four years after the initial screening. While the Gamma-GT values in both groups decreased significantly, there were a number of big differences between the two groups. In terms of absenteeism, the control group took twice as many sick days as the intervention group. In terms of hospitalisation, the control group had twice as many days in hospital as the intervention group. Finally, it was noted that over a five-year follow-up period, there were twice as many *deaths* in the control group as in the intervention group.

This study powerfully supports the view that simple intervention based on regular feedback about a blood test marker can have a significant effect on the drinking habits and physical health of a population. The implications for industry of reduced mortality, reduced absenteeism and reduced hospitalisation are obvious.

Identifying the cause of alcohol abuse

Sometimes there is no obvious underlying reason for alcohol abuse or at least not one that the individual is prepared to admit to. It is often a response to problems which arise at home and relate to marital and family difficulties, financial or legal problems, or problems with illness or bereavement. It is not uncommon for people to drink as a reaction to worries which they have about their own health. The fact that alcohol will only worsen the situation is not always apparent to them.

It is important to ask whether the excess alcohol consumption could be a response to a preventable source of stress in the workplace, such as:

- Too much or too little work
- Excessive hours and overtime
- Conflict with colleagues
- Monotony
- Too much responsibility
- Inadequate resources
- Inadequately defined job responsibilities.

If any of these factors might be causing drink problems for an employee, this is a problem for senior management.

Availability of alcohol in the workplace may be part of the problem in some cases.

Managing alcohol abuse

In general, there are three possible reactions to problem drinking in the workplace:

- *The 'head in the sand' approach* – 'if you ignore the situation it might simply go away.' Pretending it does not exist will solve nothing. The employee's drink problem is unlikely to go away and the company will continue to bear the costs of reduced output and efficiency and also the increasing risk of accidents in the workplace.
- *Dismissal.* This may have the attraction of being a quick solution to the problem but it does not, of course, do anything to help the employee. In some cases it may be a valuable and highly skilled employee who will be difficult to replace. There is also the risk of a claim for unfair dismissal.
- *Offering support and assistance with treatment.* This is the most appropriate reaction, at least in the first instance. If the problem of alcohol abuse can be dealt with in a non-threatening and supportive environment, the individual is not only more likely to want to seek treatment but is also more likely to recover and return to full and productive employment. If the outcome is successful, then obviously both the employee and the employer benefit.

Establishing a company policy on alcohol

Many companies and organisations, including British Nuclear Fuels, Hoover, Marks & Spencer and Thames Television, have implemented a formal policy on the management of alcohol abuse in the workplace. Such policies tend to differ in detail, but contain many important principles in common:

1. Whether the drinking problem comes to light with observed changes in appearance or work performance, or through the screening process, or a combination of both, it should be discussed with the employee in a non-hostile manner. The overall company policy with regard to alcohol abuse should be carefully explained. It is extremely important that the employee understands and participates in the management procedure and knows that strict confidentiality will be maintained throughout.

2. The individual should be encouraged to take advantage of specialised counselling/treatment facilities which the company will provide. Those organisations which have physicians and nurses in an occupational health department are normally able to offer appropriate counselling programmes on an in-house basis. Others need to make arrangements with specialist external agencies, including the patient's own GP. Many GPs are extremely helpful and supportive in managing alcohol problems and it is a good policy to encourage the individual to seek the help and advice of his family doctor at an early stage.

3. Where appropriate, leave of absence should be given to encourage full participation in the various counselling and treatment programmes. The employer should be kept closely informed of the individual's progress during this period and the likelihood of his returning to work.

4. Every effort should be made to ensure that the individual is able to return to the same job after the counselling/treatment programme has been completed.

individual is able to return to the same job after the counselling/treatment programme has been completed.

5. Employees who are identified as having an alcohol problem but who decline to accept referral for specialist help, or who discontinue a course of treatment before it is satisfactorily completed, are normally subject to the company's disciplinary procedures.

6. A senior person within the organisation must be given overall responsibility for the implementation of the alcohol policy. In most organisations this is the personnel director/manager. On a day-to-day basis this responsibility may well be delegated to line managers and supervisors who have the main responsibility for identifying employees with problems.

7. Whenever possible the full co-operation and assistance of the trade unions should be sought. This considerably enhances the acceptance of the policy among the workforce and greatly increases its chances of success.

8. The contents of the policy must be adequately communicated to *all* members of the workforce. Employees need to know that they can seek treatment, help and advice in an atmosphere that will be sympathetic and supportive.

Companies vary so much in size, geographical location, age and sex distribution and social attitudes that a 'blueprint' alcohol policy cannot be given here. Below is an outline policy which can be fleshed out according to individual requirements. There are a number of organisations that will assist in the drawing up of a policy, and these are listed in the Appendix.

Outline of an alcohol policy

1. Recognition

The company should clearly identify alcohol abuse as a growing problem in the community and express its concern that the company's employees should not damage their health, their social and family relations or their employment prospects by the inappropriate use of alcohol. For this reason, the company has formulated a specific policy aimed at educating the workforce as to the potential hazards involved in excess alcohol consumption, and also to provide advice and support for those employees who have alcohol-related problems.

2. Education

The policy should be emphatic in its promotion of education and information as the essential components of a strategy for preventing alcohol abuse. The educational programme should include information on the scale of the problem; the physical and psychological consequences of alcohol abuse; the concept of 'units' of alcohol; advice on safe drinking limits; and the potential dangers of alcohol in the workplace, especially with reference to increased risk of injuries and accidents. It should also emphasise that turning a blind eye to the problem is an inappropriate response, which may jeopardise not only the individual concerned but also his work colleagues.

3. Responsibilities

The document should clearly state who has overall responsibility for the implementation of the policy and should also identify those individuals or groups to whom that responsibility may be delegated.

4. Further training

Senior managers, supervisors and others will be responsible for the everyday implementation of the policy and they will clearly require further training. The nature, duration and content of that training should be clearly stated.

5. Identification

The policy should indicate that the identification of the problem drinker is not always easy, but that some indicators may be a deterioration in work performance, increased absenteeism, poor concentration and a deterioration in personal appearance. This means an increased cost for both employer and employee; the one in loss of production and the other in lost pay. The attitude of the company towards the use of health screening in both the pre- and post-employment situation should be clearly set out.

6. Management attitudes

The policy should emphasise the desire of the management to promote the general wellbeing of all employees and to provide, so far as possible, a working environment which is conducive to that wellbeing. In the case of employees who develop alcohol-related problems, the policy should emphasise the commitment on the part of management to provide every possible support and assistance to the employee in dealing with the problem.

7. Employee assistance

The document should clearly identify the initial contact point for any employee with an alcohol-related problem. The facilities and the arrangements for counselling and/or treatment, whether 'in-house' or via external agencies, should be clearly set out. In relation to the treatment or counselling period, the employee should be reassured that all benefits and rights laid down in the contract of employment would be safeguarded.

A guarantee of absolute CONFIDENTIALITY must be given.

8. Leave of absence

The policy should indicate that absence from work due to

treatment/counselling for alcohol-related problems, will be taken through the normal sick leave entitlement.

9. Return to work

The policy should indicate that, so far as possible, the company will endeavour to ensure that an employee will be given an opportunity to return to his/her old job, on return to work.

10. Disciplinary procedure

The policy should make it absolutely clear that while the company will assume a sympathic and understanding position with regard to employee alcohol-related problems, the organisation also expects that individuals will be prepared to help themselves. In situations where an employee refuses to accept an offer of help, or who discontinues a course of treatment and then reverts to previous unsatisfactory levels of conduct or performance, then the matter will be dealt with by the agreed disciplinary procedures. It should also indicate that any employee who receives help in the form of specialist treatment, but whose conduct or work performance deteriorates subsequently, will require to have the new situation considered on its merits. In some cases a further course of treatment may be thought appropriate, but, equally well, in some cases the disciplinary procedure may need to be invoked.

DRUG ABUSE

Much of what has been said about managing alcohol problems, also applies to the management of drug abuse; indeed, many companies have developed a combined policy for alcohol and drugs which seems sensible. The scale of alcohol abuse is much greater and drug abuse has serious legal aspects, but the basic approach is very similar: education; the identification of the drug abuser; and the development of a company policy to deal with issues of drugs in the workplace.

Educating the workforce about drugs

As with alcohol, the need for education about the potential dangers of drug abuse cannot be overemphasised. The educational programme should include information about the physical and psychological consequences of drug abuse; the increased risk of accidents and injuries in the workplace; the association of the AIDS virus with the use of intravenous drugs, and the serious legal implications of using and supplying drugs.

Identifying the drug user

Identification of the drug user, like the problem drinker, may be a matter of observing appearance and work record, or it may be done by specific screening techniques.

Drug abuse, like alcohol abuse, results in increased absenteeism, deteriorating work performance, impaired concentration, unreliability and an increased risk of accidents and injuries. It is sometimes said that drug abusers can be identified by specific physical characteristics such as sweating, irritability, muscle tremor, and dilation or constriction of the pupils. Such signs are significant but they are fairly non-specific – some of them may equally well occur with various forms of physical illness and, of course, in alcohol abuse. Indeed, it should never be forgotten that drug abuse and alcohol abuse may well occur together.

Screening for drug abuse in this country is still a contentious issue. In the USA, where the drug problem is significantly greater, routine testing for drug abuse is now commonplace and about half of the Fortune 500 companies regularly undertake some form of drug screening. These tests are usually carried out as part of a normal pre-employment medical examination, or as part of screening programmes and periodic medical examinations for those already employed. Some companies in America even carry out 'spot testing', and individuals who test positive are usually dismissed immediately.

There are two commonly voiced objections to routine drug testing in this country. First, it is claimed, the laboratory techniques used to measure drug use are not reliable. This could mean an employee being dismissed following a positive result, when in fact the test was at fault. Given the extraordinary sophistication of modern laboratory technology and the fact that any positive test would be repeated before reporting, it seems difficult to sustain this argument.

The other argument (often voiced by trade unions) is more of an industrial relations one: that screening an entire workforce might reveal one or two employees with a drug problem but, in the process of trying to identify these individuals, the company would alienate the remainder. This argument is more sustainable because, even though little hard information exists, the prevalence of drug abuse does appear to be significantly less than that of alcohol abuse. One study carried out by British Nuclear Fuels revealed that less than 0.5 per cent of employees had evidence of a drug-related problem, while around 300 employees at one site showed biochemical evidence of alcohol abuse. So it would seem much more rational to carry out screening programmes for alcohol abuse than for drug abuse.

But, of course, we have to bear in mind that the drug problem in the USA is very different to that which exists in the UK. If evidence suggests that the problem is much more widespread than we have imagined, then the argument for routine drug testing becomes more sustainable, particularly in the light of the legal implications. At the present time, unless a company employs a population which is in a particularly high risk group (and it is not easy to imagine what this might be), then routine testing for substances of abuse other than alcohol is probably not warranted.

A drug policy

A number of organisations have combined their approaches to problem drinking and drug abuse and have, sensibly enough, devised one policy to cover both. When the drug policy has been developed, it must be implemented. There is no point in having a document which has only been half communicated to the management or to the supervisors, since this may encourage them to turn a blind eye to clear cases of drug misuse, in an attempt to protect the individual involved. As we have seen, this is not a solution since it will end in misery for the employee and increased costs for the employer. Moreover, ignoring a known case of drug abuse may have significant legal consequences.

As with alcohol abuse, the main aim of the policy is to take the problem out of the disciplinary procedure and seek to create an environment in which an individual can seek treatment and counselling while maintaining continuity of employment. A drug policy should follow the same principles set out for the alcohol policy, with the addition of a section on the legal implications of drug abuse (the Misuse of Drugs Act 1971). Particular attention should be paid to the questions of possession and allowing premises to be used, which are dealt with in sections 5 and 8 of the Act respectively.

CHAPTER 12

PREVENTING CANCER

In 1981 two of our most distinguished scientists, Doll and Peto, produced a major review, 'The Causes of Cancer', for the American Congressional Office of Technology Assessment. This report emphasised the importance of potentially avoidable lifestyle factors in the origin of many forms of cancer (see Table 12.1). One of its conclusions was that between 80 and 90 per cent of all cancers could, in principle, be avoided. Moreover, the authors believe that in America (the country for which the report was prepared), about two thirds of all cancer deaths are caused by tobacco and diet, a point well illustrated in Table 12.2.

It should be stressed that the importance of these risk factors to each of the cancers mentioned varies considerably. In the case of lung cancer, for example, the association with cigarette smoking has been so clearly demonstrated that we can reasonably regard cigarette smoking as the *cause* of lung cancer. But in many other cases the contribution of such factors may be much less certain; for example, the relationship between alcohol intake and breast cancer. It is also important to bear in mind that these are not the only factors involved. In the case of breast cancer we need to take into account the possible contribution of the contraceptive pill and also a genetic contribution.

Primary prevention

In primary prevention the aim is education about principal risk factors for the various forms of cancer, in the hope that appropriate lifestyle changes will reduce overall risk to a minimum. The need for intensive health education is a consistent

Table 12.1 *Cancer deaths and risk factors*

	Cancer site	No of deaths in 1986 and percentage of total cancer deaths (%)		Risk factors
Males and Females	Lung	40,142	(25.3%)	Tobacco smoking
	Mouth	445	(0.28%)	Tobacco smoking, alcohol
	Oesophagus (gullet)	5,227	(3.29%)	Tobacco smoking, alcohol
	Stomach	10,918	(6.87%)	Tobacco smoking, alcohol, diet
	Pancreas	6,776	(4.26%)	Tobacco smoking, alcohol
	Liver	1,450	(0.91%)	Alcohol, diet
	Bowel (rectum and colon)	19,081	(12.01%)	Diet (high fat, low fibre), obesity
	Bladder	5,238	(3.29%)	Tobacco smoking
	Breast	15,366	(9.67%)	Alcohol, diet, obesity, stress, contraceptive pill
Females	Cervix	2,234	(1.4%)	Tobacco smoking
Males	Prostate	7,574	(4.77%)	Obesity
	Total no of deaths	114,451	(72%)	
	All other cancers	44,318	(28%)	
	All cancer deaths	158,769	(100%)	

Table 12.2 *Causes of cancer*

Cause	Estimated percentage of all cancer deaths (%)
Tobacco	30
Alcohol	3
Diet	35
Food additives	Less than 1
Occupation	4
Pollution	2
Industrial products	Less than 1
Medicines and medical procedures	1
Infection	10
Other and unknown	13

Data from 'The Causes of Cancer' (OUP 1981) by Doll and Peto

theme in this book because of its fundamental importance as the cornerstone of any strategy for preventing disease and improving the health of a population. While it is true that in the case of cancer we do not yet have all the answers, we certainly have sufficient evidence to encourage appropriate action in relation to a number of lifestyle factors such as cigarette smoking, alcohol consumption, obesity, and various components of our diet. Thus, once again, the first and most important step a company can take in cancer prevention is to *educate* the workforce.

Health education has been discussed elsewhere in the book, so here we concentrate on secondary prevention, that is, the use of screening techniques to identify disease which is already present. It is believed that early detection gives a much greater chance of cure and improves long-term survival rates. We shall discuss cancer screening techniques in relation to seven common forms of cancer: lung; breast; cervix; ovary; bowel; prostate and testicular cancer.

Lung cancer

Lung cancer is the most common form of cancer in the United

Kingdom and accounts for about 40,000 deaths each year (28,500 males and 11,500 females). This means that, on average, at least 100 people die from lung cancer every day in the UK. It is a particularly virulent and dangerous form of cancer because, irrespective of treatment, only about 7 per cent of all individuals who develop it are alive after five years. Given that the overwhelming majority of lung cancer (well over 90 per cent) is caused by tobacco smoking it is, by definition, almost completely avoidable.

The normal way of diagnosis is by a standard chest x-ray, but this is not a screening technique. Screening implies that there is some *benefit* for the patient, and the use of x-ray in this context is of doubtful value since, by the time the tumour shows up on a chest x-ray, the outlook is already extremely poor. It might be argued that there is some benefit for the patient in that he or she is given an early diagnosis. But this is definitely not an advantage in all cases – many patients prefer not to know until it is absolutely necessary.

The poor outlook for lung cancer patients emphasises the need for prevention of this disease, and the most effective 'screening' technique is therefore to identify cigarette smokers and do everything possible to get them to give up.

Breast cancer

Britain has the highest rate of breast cancer in the world. Each year in the UK about 15,000 women die from breast cancer and about 25,000 new cases are diagnosed. It is the most common type of cancer in women, accounting for about 20 per cent of all new cases, and it is estimated that about one in twelve women will develop breast cancer at some time in their life. For women between the ages of 35 and 54 years, breast cancer is the most common single cause of death.

By contrast, breast cancer in males is a rare disease and causes about 120 deaths annually.

The majority of breast cancer cases are in older women; 80

per cent of all cases occur in women over the age of 50, and almost 50 per cent in women over the age of 65. Both the incidence of the disease and mortality rates are increasing, although these increases are seen mainly in the older age groups.

Throughout the world, over half a million women develop breast cancer each year, but no less than half of all these cases occur in North America and Europe which contain less than one fifth of the female world population. This is compelling evidence to suggest that breast cancer is a disease which occurs mainly in Western developed countries. Although there is considerable variation in mortality rates between countries, the differences seem to be due primarily to social and environmental factors, not genetic ones. As we understand more about the natural history of this disease, we should therefore be in a much better position to carry out primary prevention and rely less on early detection.

Risk factors for breast cancer

A number of factors have been identified which contribute to the risk of developing breast cancer. (See Table 12.3.) For some of these the association is clearly established; for others, the association is still being assessed.

Table 12.3 *Risk factors for breast cancer*

Risk factors for breast cancer	
Strong association	*Possible association*
• Increasing age	• High fat diet
• Late childbearing, i.e. first child at 30 years of age or older	• Hormone replacement therapy (HRT)
• No children	• Alcohol
• Early onset of menstruation	• Stress
• Late menopause	
• Family history of breast cancer (first degree relative)	
• Obesity	
• Radiation	
• Contraceptive pill	

A major research programme at present underway in London and Oxford has found good evidence that the oral contraceptive pill is a significant risk factor for breast cancer in younger women. In women who have been using the pill for up to four years, there is no increased risk of breast cancer, but beyond that point, a clear hazard emerges. Four to eight years of use brings an increase in risk of about 40 per cent rising to 70 per cent after more than eight years. The women studied were all under the age of 36.

Although these figures sound alarming, it is important to remember that breast cancer is relatively uncommon in this age group; in women under 36 it affects about one in 500. This means that even a 70 per cent increase in risk would only raise an individual's chances of developing the disease to about one in 300. That level of risk will clearly deter some women, but others may consider the risk worth taking in return for the convenience and reliability of the pill in other respects.

The research group is now carrying out a similar study to determine whether long-term pill users *remain* at an increased risk of developing breast cancer, at least up the age of 45.

It is important to recognise that there are many different factors which contribute to a woman's resistance or vulnerability to breast cancer. For example, high fat diets are known to increase the risk, whereas recent studies have suggested that breastfeeding can greatly *reduce* the risk.

Screening for breast cancer

The technique used to screen for breast cancer is called mammography, and it involves taking highly specialised x-ray pictures of breast tissue to see whether any cancerous lesions are present. Mammography is normally combined with examination of the breast ('palpation'), usually by a doctor or nurse; this improves the predictive value of the screening test. It is extremely important that women should be taught the technique of breast self-examination. Evidence shows that if breast self-examination is to be beneficial it must be practised regularly and proficiently. It is an important and useful supplement to mammography, but it can in no way be regarded as a substitute.

Is screening for breast cancer beneficial? The answer to this question is an unequivocal *Yes*. Trials in New York and Sweden have shown reductions in death rates from breast cancer of between 30 and 40 per cent achieved by screening.

Is screening safe? Some concern has been expressed in the past about the use of x-rays to identify breast cancer, since there was always a possibility that the use of radiation might actually *cause* more breast cancers than it identified. An increased incidence of breast cancer has certainly been reported after exposure to high doses of radiation, especially in atomic bomb survivors and in some patients who have received radiation as part of a therapeutic programme. But the doses used by today's modern mammographic units are so minute as to be virtually irrelevant. We do not yet know whether very low doses of radiation cause breast cancer, but if they do, the incidence is so low that it has never been observed. For all practical purposes, the risk can be disregarded.

The main disadvantages are to do with the false positives and false negatives which any screening programme inevitably produces and relate to the sensitivity and the specificity of the test. Sensitivity is the ability of the test to detect a disease. In the case of mammography, the estimated sensitivity is about 80 per cent. This means that for every ten women with breast cancer, mammography will identify eight, and two of the cases will not be identified and will be told that they do *not* have the disease. In other words, two out ten will be false negatives. The use of breast palpation in *combination* with mammography may increase the sensitivity to around 90 per cent.

Specificity is the ability of the test to exclude people who do not have the disease. Mammography has a specificity of about 90 per cent. This means that for every ten people who do *not* have disease, nine will be correctly identified as being disease-free, but one will have a mammogram reported as being abnormal, even though there is no disease present, i.e. a false positive. These individuals will require further assessment before the possibility of malignancy can be excluded, and this can obviously create considerable anxiety.

Only organisations that have appropriate professional standards and quality assurance should be involved in breast cancer screening. The recent Pritchard Report will form the basis of a comprehensive quality assurance programme for the NHS and the Royal College of Radiologists is currently considering recommendations for the private sector.

The UK National Breast Screening Programme is planned to be set up by 1990. With nearly £55 million funding, the intention is that England, Wales, Scotland and Northern Ireland should eventually have a national network of screening centres, including mobile units for rural areas.

A number of private organisations offer breast screening programmes, some of which are mobile but most of which are hospital-based. Obviously, the main advantage of the mobile service is that it enables breast screening to be carried out either at or near the workplace, making it more accessible and convenient for employees, and causing minimal disruption to the working day for the employer.

At what age should screening start? The National Breast Screening Programme will invite all women aged 50 to 64 to attend. Women aged 65 and over will not be routinely called for screening, but will be welcomed if they request it.

The rationale for this is partly that breast cancer is primarily a disease of older women and partly that trials carried out so far have not yielded any conclusive evidence that screening women *under* the age of 50, confers any benefit. Furthermore, some radiologists believe that the nature of breast tissue in younger women makes interpretation of mammograms substantially more difficult and therefore raises the problem of a greater number of false positives and false negatives.

However, many authorities believe that breast cancer screening should be extended to younger women. They argue that there are studies available to suggest significant benefit from breast screening in younger women, and while accepting the difficulties of radiological interpretation of the mammograms, they do not accept that these problems are insurmountable. Furthermore, they argue very strongly that while it is true that most women who develop breast cancer are over the age of 50, a very significant proportion of cases occur in younger women. For example, of the 21,300 or so women who developed breast cancer in England and Wales in 1984, 4300 (20 per cent) were in women *under* the age of 50. Moreover, since breast cancer in younger women tends to be much more aggressive, many people believe that it is wrong to deprive them of the opportunity for early diagnosis. In keeping with this view the American College of Radiologists has recommended that the first or baseline mammographic examination should be performed between the ages of 35 and 40 and this is the advice adopted by most private screening organisations.

What should be the screening interval? The National Breast Screening Programme will carry out screening on a three-yearly basis, although some private organisations recommend screening every one or two years.

The need for education

As with other forms of screening, one big problem is that those who most need it because they are in a high risk group, are often the ones who never bother to attend. This is true even when the service is free of charge and even when it is delivered on site. The reasons for non-attendance are complex, but mainly originate in a surprising level of ignorance about the nature of breast cancer and the importance of regular screening. Many women still believe that a diagnosis of breast cancer is nothing short of a death sentence. But the survival figures for England and Wales show that, on average, 64 per cent of women diagnosed with breast cancer in 1979 were alive five years later. The really critical issue is the time at which the cancer is identified and treated. The more advanced the cancer, the poorer the outlook; if the cancer is caught at a very early stage, the prognosis in many cases is excellent. Table 12.4 shows five-year survival rates for each of four tumour stages. Stage 1 is the earliest and Stage 4 the latest, when the disease has advanced with spread to other organs.

Table 12.4 *Breast cancer survival rates for four tumour stages*

Stage	5-year survival rate (%)
Stage 1	84
Stage 2	71
Stage 3	48
Stage 4	18

So the message is very clear. The earlier the diagnosis is made the better are the prospects for a full recovery; and the best means of identifying breast cancer at an early stage is by regular screening involving self-examination and mammography.

There is a wealth of information available on the subject of breast cancer including several videos, books, and other literature. The employer can do an enormous amount to raise the general level of awareness of the issues by regular distribution of this information and by encouraging women to take full advantage of whatever screening services are available, whether they are provided by the NHS or by the private sector. After all, no matter how comprehensive and how sophisticated screening techniques become, they will be completely ineffective if those women who most need them do not avail themselves of the service.

Cervical cancer

This is a specific form of cancer affecting the cervix – the lower part, or neck, of the uterus (womb). In the UK in 1986 there were 4567 new cases of cervical cancer and 2234 deaths. The figures are tragic enough, but much more so when one considers that cervical cancer is an almost totally preventable disease. The majority of women dying from cervical cancer do so simply because they have never had a cervical smear. Indeed, about 80 per cent of women who develop cervical cancer are over the age of 40, and most of them have never had a cervical smear test. However, since the mid-1960s, there has been a tenfold increase in the incidence of cervical cancer in women under the age of 40 and many authorities now believe that we are on the verge of an epidemic of cervical cancer in younger women.

Figure 12.1 shows the uterus which is shaped like an upside-down pear, the wider part of which is called the body and the narrow part the cervix. Seen face on, the cervix has an oval appearance and a central hole. In a woman who has not had children, this central hole is round, whereas in a woman who has had one child or more it tends to be more slit-shaped. It is from this external part of the cervix that a 'smear' is normally taken.

It has been known for over a century that cervical cancer is a sexually transmitted disease. It follows that the more sexual partners a woman has, the greater the degree of risk. But while it is true that promiscuity significantly increases the risk of cervical cancer (prostitutes have a very high rate) it does *not* follow that all women with cervical cancer are promiscuous. It is fundamentally important to make this point because many women are reluctant to come forward for cervical smears for fear that if anything is found to be abnormal, they will be labelled promiscuous. Moreover, it is not merely promiscuity in the woman herself which affects her vulnerability, it is also that of her partner or partners. Thus, the more sexual partners a man has, the greater the risk for the woman.

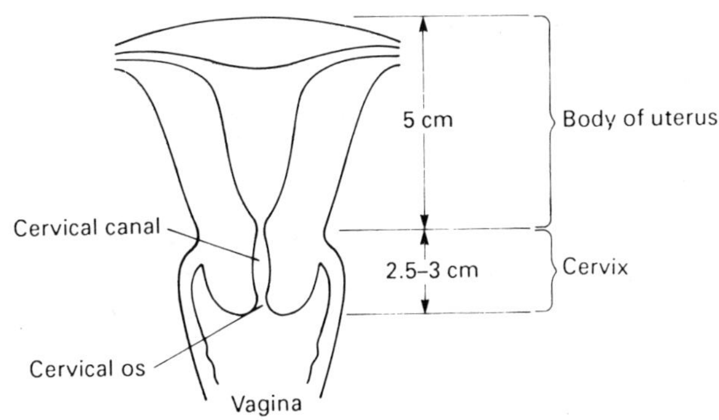

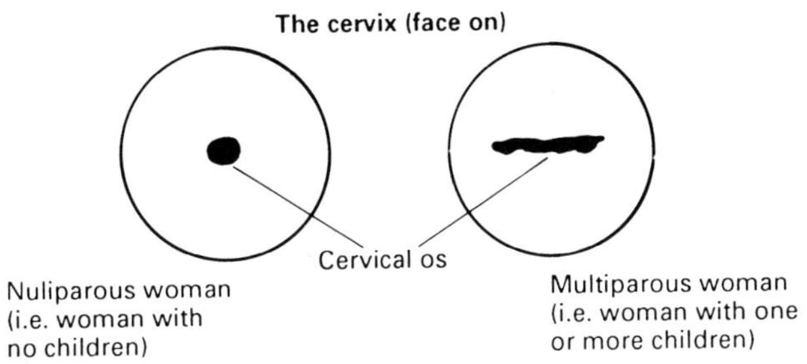

Fig. 12.1 *The uterus and cervix*

These facts strongly suggest that some factor is passed from the male to the female which increases the risk of cervical cancer. There is a growing body of evidence to implicate a special form of virus known as the human papilloma virus (HPV), or wart virus. About 90 per cent of women with cervical cancer have antibodies to HPV, indicating that they are infected with this virus. It seems clear from all the evidence that males are the primary carriers of HPV, and that this virus is passed on to the woman during intercourse. Therefore, the presence of genital warts in the male significantly increases the risk of cervical cancer for the female. However, a man may be a carrier of HPV without showing any sign of it.

It might be expected from this that all women with evidence of HPV infection (with or without genital warts), would develop cervical cancer. This is clearly not the case. So what are the additional risk factors which act in concert with HPV infection to produce cancer of the cervix? Heavy cigarette smoking is known to be a strong risk factor for cervical cancer. Heavy smokers in particular (over 20 cigarettes a day) may be up to seven times more likely to develop cervical cancer than non-smokers. The mechanism for this is not clear, but it seems that the body's immune system is adversely affected by nicotine, and this may make the cervical cells much more susceptible to the potential effects of the wart virus.

The risk of developing cervical cancer also seems to be influenced by the type of contraception used. Barrier methods such as the sheath and the diaphram appear to have some protective effect. This is presumably because they prevent the male semen (which is known to carry HPV) from coming into contact with the cervix. There is also some evidence, although by no means conclusive, that the contraceptive pill may slightly increase the risk of cervical cancer.

Another important factor to take into account is the number of children a woman has and also her age at the first pregnancy. The risk of cervical abnormalities and cervical cancer increases with the number of pregnancies. And the younger the woman is at the time of her first pregnancy, the greater the risk appears to be. Again, the reason for this is not entirely clear, but it may well relate to the fact that during pregnancy and childbirth, the cells covering the cervix change and develop at a much more rapid rate than normal, thus making them more vulnerable to other agents such as the HPV virus. It is also possible that trauma to the cervix during childbirth has a part to play in promoting this increased risk.

Finally, some evidence has recently emerged that some women may be genetically more at risk of developing cervical cancer than others. Although this is likely to be a fruitful area of research in the future, for the moment, it has no practical applications. Table 12.5 summarises the risk factors for cervical cancer.

Table 12.5 *Risk factors for cervical cancer*

Risk factors for cervical cancer
• Sexual behaviour, i.e. the more partners, the greater the risk • Sexual behaviour of the male, i.e. the more contacts, the greater the risk to the woman • Presence of HPV infection in males and females ± genital warts • Cigarette smoking • Age at first pregnancy, i.e. the younger the first pregnancy, the greater the risk • The number of pregnancies, i.e. the more pregnancies, the greater the risk • Oral contraceptive pill, i.e. may increase the risk slightly • Genetic factors

Preventing cervical cancer

While the exact cause of cervical cancer remains unclear, primary prevention is difficult and in some respects may not be socially acceptable. But women should at least be aware of the fact that the more sexual partners they have, the greater the risk. They should also be aware that that barrier contraceptive methods are protective whereas long-term use of the may increase risk. Since cigarette smoking is known to be a significant independent risk factor for cancer of the cervix, this is yet another reason why all women should be advised to stop smoking.

The most effective technique for the prevention of cervical cancer currently available is the cervical smear (also called a Pap test or Papanicolaou smear).

Most cases of cancer of the cervix appear to be preceded by a long period of changes in the cells on the surface of the cervix. The aim of a cervical smear is to detect these abnormal cells *before* they have progressed to a point where they are cancerous. This is an extremely important point to understand. The aim of the cervical smear is not to identify cancer but to identify changes in the cells on the surface of the cervix *before* cancer has occurred.

The smear itself is taken by introducing a speculum into the vagina and using a spatula made of wood or plastic to take a scraping of cells from the surface of the cervix. The cells obtained are smeared on to a glass slide and then fixed in a special solution. The smear is then sent to the laboratory for analysis. According to the detailed appearance of the cells, the smear will be classified as either normal (negative smear) or abnormal (positive smear).

The abnormal smear is further classified according to the severity of changes demonstrated in the cells. The normal system used is the CIN classification (Cervical Intra-epithelial Neoplasia). (See Table 12.6.)

Table 12.6 *Cervical Intra-epithelial Neoplasia (CIN) classification of cervical smear abnormalities*

Report	Explanation and action
CIN 1	Minor abnormalities in cervical cells but may regress to normal in many cases. Usually advised to repeat smear in 6 months.
CIN 2	Abnormal smear with more advanced cellular changes than in CIN 1. Usually repeat smear in 3 months but colposcopy may also be advised.
CIN 3	Abnormal smear with definite changes of premalignancy which may progress to cancer if no action taken. Recommend colposcopy and biopsy

Women with mild or moderate degrees of change (CIN1 and CIN2) will be advised to have the smear repeated, usually after a three- to six-month interval depending upon the circumstances. Women with more severe degrees of abnormalities (CIN3) will definitely require further examination and investigations.

If a definite abnormality is found, there are a number of possible treatments which are in common use. The important point to make is that provided these abnormalities are dealt with soon enough, the problem can be cleared up completely.

Who should be screened? The simple answer to this question is that all women who are or who have been sexually active should undergo regular screening for cervical cancer. Ideally, screening should commence within two years of a woman becoming sexually active. The actual age itself is irrelevant.

How often should screening be carried out? The usual view is that women should have a smear carried out every three to five years, but there are many experts who believe that this is nowhere near frequent enough, and that the screening should be made available on an *annual* basis. Given that the NHS screening system is unable to cope with the volume of smears which it has at present, it seems extremely unlikely that it will be able to cope with a more frequent screening programme. Attempts are currently being made to develop a nationally co-ordinated cervical cancer screening programme supported by a complex computer network including an effective recall system. The difficulties involved in this are formidable and it is likely to be a long time before women in the UK are provided with the sort of screening programme which has been available in other countries for many years and which is so desperately needed in the UK.

In the ideal situation, a smear would be carried out on an annual basis, but a reasonable compromise would be to carry out annual smears until there are two negative ones in a row, and then to carry out two-yearly smears thereafter. It is important to note, however, that any women who has had an abnormal smear or who has evidence of HPV infection, should have smears performed on an *annual* basis. As a rough guide, compared with a woman who has never had a smear, the woman who has yearly smears is about 12 times better protected, while the woman who has five-yearly smears is only about 3 times better protected.

Is screening effective? There is no serious doubt among experts that properly organised screening programmes can lead to a major reduction in cervical cancer. Screening programmes have been running in various parts of the world for more than 30 years and in many countries have been extremely successful. The United States, Canada, Iceland, Sweden, Finland and the province of British Columbia have all reported significant, sometimes dramatic falls in death rates from cancer of the cervix.

We discussed the problem of false positives and false negatives in some detail in relation to breast cancer, and a similar problem exists in screening for cervical cancer. The false negative rate – that is, telling a woman that her smear is normal when in fact it is abnormal – has been a particular problem. During the past 20 years, the false negative rates for cytology screening have been reported to range from as low as 6 per cent, to as high as 55 per cent by various authorities. The Mayo clinic in the United States reported a 20 per cent false negative rate, 62 per cent of which was due to sampling error, 15 per cent to laboratory error and 23 per cent to errors in interpretation. Even in the best programmes available, false negatives still occur, and this another reason why annual screening is to be preferred.

A common problem at the present time, is the long time women have to wait for the result of their cervical smear to come through. This creates considerable alarm and distress among many women, some of whom have to wait for months. The implications of this are obvious. By the time a smear report indicating a minor abnormality is delivered to the patient, the problem may have got worse. The need for prompt and efficient reporting of results cannot be over-emphasised.

Studies carried out on women who chose not to attend for screening, are revealing. Some women said that they thought the test would be painful; others said that they would be embarrassed about having an internal examination performed; others said that they were frightened about what the result might be; some said that they did not wish to see a male doctor for this examination. One study, carried out in Manchester, showed that many of the women did not attend because they believed that the family doctor had invited them to come along for the test, because they already had cancer and the doctor knew they had it. Another group said that they believed that cervical cancer was a problem on a national basis, but that it was not a particular problem in Manchester!

What is clear from all this is that there is a tremendous educational job to be done and that we have to find more effective ways of encouraging women to have the test. In countries such as Sweden and Finland, compliance rates are as high as 70 or 80 per cent, and there is no reason why we should not be able to achieve a similar percentage in this country, if we make an effective educational campaign, supported by a well-co-ordinated and efficient screening programme, an absolute priority.

Again, the most effective contribution the corporate organisation can make, is to raise the general level of awareness of the need for regular cervical cancer screening. Recognising the inadequacies of the present system, many employers have decided to offer cervical cancer screening programmes to their employees free of charge and these services are normally carried out by private screening organisations supported by independent laboratories. The screening progamme is sometimes carried out in a mobile unit, or sometimes on the company premises. The need for absolute confidentiality and privacy if this is to be done, hardly needs stating.

Cancer of the ovary

In 1986 in the UK there were about 5000 new cases of ovarian cancer, and just over 4300 deaths. It is a serious form of cancer and only about 28 per cent of those diagnosed will be alive after a five-year period.

Currently, there is no satisfactory screening technique in general use. The one which has attracted most attention in recent years has been that of diagnostic ultrasound. Unfortunately, there are major technical and interpretive difficulties with this technique, particularly an extremely high false positive rate, and, for the moment, this would seem to preclude its use on a widespread basis.

A newer technique is to use something called a 'tumour marker'. Some forms of cancer produce specific chemical substances which can sometimes be identified by using complex immunological blood investigations. A variety of such markers now exist and it is likely that they will be of considerable use in the identification of a variety of cancers, including ovarian cancer, breast cancer, bowel cancer, lung cancer and testicular cancer.

The combination of a blood test and ultrasound may provide a much higher degree of predictive accuracy than the use of either technique alone, but this remains to be tested.

Bowel cancer (cancer of the rectum and colon)

In the UK in 1986 there were approximately 28,000 new cases of cancer of the rectum and colon, and 20,000 deaths.

In order to screen large numbers of individuals, a simple and inexpensive screening test is required. Questionnaires have been used, but they are not very specific and therefore lack predictive value. Endoscopic screening using a rigid or flexible sigmoidoscope (an instrument for examining the inside of the rectum or colon through the anus) has been widely used in the United States but not in the UK, and it seems unlikely that this will ever achieve widespread acceptance in the screening of large populations.

The most commonly used technique in screening for large bowel cancer is the occult blood test (occult simply means hidden) which is peformed on stool samples collected over a two- or three-day period.

A newer, although more expensive, test method dispenses with the need to collect the stool sample and can be used by the patient at home. After a bowel motion has occurred, the test simply involves dropping a specially impregnated piece of paper into the toilet bowl and seeing whether any colour change occurs. This test has a number of potential problems, including a significant false positive rate.

Cancer of the bowel tends to occur in older age groups, and so the generally accepted view is that screening should start at the age of 50. However, cancer of the bowel may occur in younger people such as those with a family history of the disease.

It is important to understand that a positive faecal occult blood test, does not mean that the patient necessarily has cancer. Indeed, the vast majority of patients have non-malignant causes of bleeding, such as rectal polyps or adenomas, which can be quickly and effectively treated. Since adenomas themselves may well progress to invasive cancers, the identification and treatment of these lesions is extremely important. Research carried out in Nottingham is interesting. Over about five years about 30,000 individuals in the Nottingham area, aged 50 to 75 years, were offered faecal occult blood tests. About 14,000 completed the test, 410 (3 per cent) of whom were found to have positive results and were subsequently investigated. Cancer was detected in 33 patients and adenomas were found in 136 patients.

A special immunological blood test for bowel cancer has also been developed, known as CA195. It is likely that this will receive increasing attention not only as a screening technique, but also as a means of evaluating the treatment of various forms of bowel cancer over a period of time.

Screening for colorectal cancer by the faecal occult blood method, is gaining acceptance as an important means of detecting early stage colorectal cancer. The detection and removal of potentially malignant polyps (adenomas) may well reduce the incidence of colorectal cancer in a screened population and this, together with the earlier detection of malignant tumours, offers the hope that deaths from colorectal cancer may eventually be reduced. The question as to whether or not early detection of bowel cancer will *necessarily* lead to an improvement in long-term survival, has yet to be determined. Logic would certainly suggest that this is the case and early results from a number of trials are encouraging.

Prostatic cancer

Cancer of the prostate in males is a common condition and in 1986 there were just over 7500 deaths from this condition. It is particularly tragic because unlike many other types of cancer, there may actually be very little in terms of warning signs. Indeed, many patients present with symptoms related to the spread of the cancer into other organs, rather than to the prostatic cancer itself. Successful treatment depends on finding the cancer at an early stage, but early detection of prostatic cancer has been an ellusive goal. The normal method of screening is a rectal examination, sometimes used in conjunction with the measurement of acid phosphatase in the blood. Unfortunately, neither of these techniques is very specific and they have a low predictive value.

A number of newer techniques have recently become available which make the early diagnosis of prostatic cancer much more feasible. The use of an immunological blood marker, known as prostate specific antigen (PSA) has been shown in a number of studies to be useful marker of prostatic cancer in the general population. The use of prostatic ultrasound, incorporating a rectal probe, also gives much valuable information as to the presence or absence of significant prostatic abnormalities. The combination of these two techniques may well prove to be extremely useful in men aged 55 and over.

The limiting factor in screening for prostatic cancer is the expense involved. It is likely to be made available only through private screening organisations rather than the NHS.

Testicular cancer

Testicular cancer is an uncommon condition, and in the UK in 1986 accounted for 146 deaths. However, evidence suggests that it is on the increase, primarily because of an increase in cryptorchism (undescended testes). Cryptorchism is a major risk factor for testicular cancer, with about 1 in 50 children with undescended testes going on to develop a malignancy. If the cancer is caught at an early stage, it is completely treatable. Testicular self-examination should be for men exactly what breast examination is for women. It is therefore important that all males should be given advice as to how to screen themselves for this condition.

Summary

Cancer in the workplace should be considered in terms of primary and secondary prevention.

Primary prevention involves educating the workforce as to the major risk factors involved in producing various types of cancer. Advice to stop cigarette smoking, reduce alcohol consumption, include sufficient fibre in the diet and so on are all important, and for certain forms of cancer such as lung cancer, primary prevention may virtually eliminate the risk.

Secondary prevention by screening is available for many forms of cancer. All women should be strongly encouraged to participate in breast cancer and cervical cancer screening programmes when they are made available, and the corporate organisation can do a great deal to emphasise the importance of this to female employees. The National Breast Screening Programme will be made available to all women aged 50 to 64, but not to women from younger age groups. A well co-ordinated and integrated national screening programme for cervical cancer is also envisaged, but this is still a long way from full implementation.

Many private organisations offer screening for various forms of cancer, as a hospital-based programme, as a mobile service or actually on site. Whether an organisation wishes to take advantage of these services will depend upon the resources available and logistic considerations. If such a programme is introduced, it is vital that employees are fully briefed *prior* to the programme commencing in order to ensure the maximum take-up rate. Issues of confidentiality and privacy are central to the success of such programmes.

It is appropriate to leave the last word in this chapter to Professor Sir Richard Doll to whose work we referred at the beginning.

> On a worldwide scale the differences in incidence that have been observed encourage the belief that all the common types of cancer are largely avoidable, in the sense that it should be possible to reduce the risk of developing each type by at least a half and often by 80 per cent or more.

CHAPTER 13

MANAGING STRESS

'. . . why do I yield to that suggestion
whose horrid image doth unfix my hair
And makes my seated heart knock at
my ribs against the use of nature?
Present fears are less than horrible imaginings'.

Macbeth

In Part 2 we discussed the origins of stress at three levels i.e. the 'macro' or global, organisational and individual. These are summarised in Table 13.1. Of course the levels are not as clearly defined as the table implies, and there is an enormous amount of overlap and interaction. In this chapter we shall look at stress management strategies at an organisational and individual level. The macro sources of stress are put aside for present purposes because there is not very much we can do about them. We all have to accept that at a certain level there are stressors in our lives over which we can have little or no influence. But the important thing to remember is that even if we cannot change the circumstances themselves, there is a great deal that we can do to influence our reactions, attitudes and ability to cope.

Organisational stress management strategies

The organisational approach to stress management is based upon the principle of a fundamental respect for the individual human being at work. This statement may seem trite and obvious to some, but there are a great many managers who do not believe in the principle, and a great many more who may believe it but do not act accordingly. To manage stress

Table 13.1 *Sources of stress*

Stressors (source)	Causes
'Macro'/global	Major socio-economic change (e.g. Black Monday) New government legislation Political developments and trends Environmental threats Constant change
Organisational/corporate	Poor communications Changes in technology (e.g. information technology) Mergers, takeovers, company restructuring Threat of unemployment, redundancy Company liquidation Unrealistic demands, expectations Lack of adequate resources, training Insensitivity to employee opinion Lack of direction and leadership Lack of managerial support Poorly defined goals, responsibilities
Individual	Family problems, pressures Financial problems Ageing Illness Self-doubt Perceived lack of skills Change of career

within the organisation we have to recognise that we are dealing with individual people who should be respected and treated as such.

At this point it is worth reminding ourselves as to the reasons why some form of stress management within the organisation is important.

- By reducing efficiency, effectiveness and therefore productivity, stress may transform a good employee into a bad one.
- Undue stress may contribute to accidents and mistakes which can be extremely costly.
- Stress may destroy morale and motivation.
- Stress may contribute to alcohol abuse and other forms of drug abuse.

- Stress may contribute significantly to various forms of mental and physical disease.
- Stress alone may cause significant levels of absenteeism.

So what important points should senior management bear in mind when attempting to offset these potentially damaging effects? There are 12 major components which together form the basis of a highly effective strategy for managing adverse organisational stress. Attention to all of these elements is essential for success. In essence, they amount to little more than a set of principles for good management practice and will therefore be familiar to most executives reading this book. But given that stress-related disorders appear to be on the increase throughout our corporate organisations, there would appear to be a fairly widespread inability to put these principles into practice. All too often individuals are labelled as having some form of stress-related problem, when in fact they are nothing more than victims of bad management.

We are all familiar with the syndrome. An individual is promoted to a position which is beyond his ability and not commensurate with his experience. He then finds that the additional training provided is completely inadequate, and on top of that discovers that the position is grossly under-resourced. Of course he shows signs of distress. Sending him to a stress counsellor or an occupational psychologist is not only pointless, it is potentially dangerous. He is being asked to adjust to a situation for which he is clearly unsuited and in which he may be exposed to an unacceptably high level of risk. This is not only bad for the individual but clearly could damage the organisation. Unfortunately, the structure of many corporate organisations in this country does not allow an individual to admit that he is actually unable to perform a job properly, and senior management cannot admit to having made a mistake in selecting that individual in the first place. The emperor has no clothes, but no one is saying anything!

We shall now discuss 12 basic principles which together

form the basis of a strategy for managing stress within the organisation.

1. Awareness

The first principle, which may seem rather obvious, is that senior management needs to be *aware* that excess stress within the workforce can be immensely damaging to the success of the organisation. Many managers, when presented with hard evidence of the damaging effects of stress on company performance, are genuinely surprised because it had never actually occurred to them that the problem existed. Others adopt a 'head in the sand' approach which says that stress-related problems do exist, but not within their own organisation. A sensitivity to these issues should at least ensure that managers are able to respond appropriately and promptly when undue levels of stress within the company begin to threaten its performance.

The importance of awareness and willingness to respond is well illustrated by the experience of the Abbey National Building Society. Between 1982 and 1987 the society experienced a sharp rise in the number of armed raids, and at one time was experiencing up to two raids each week. Some branches were even robbed two or three times. The predictable effect of all this was that stress levels within the workforce rose rapidly and staff morale deteriorated in consequence. At first, the senior managers within the organisation were not sure how to respond and as a result there was a serious lack of co-ordination and organisation. Recognition that the problem existed enabled them eventually to develop specific strategies to improve security and also to give appropriate support to those employees who had been victims of armed raids. The result of their efforts was substantially improved morale and performance and improved security throughout the company. First of all by being aware, and secondly by listening to what the employees had to say, the company was able to develop a specific response to a set of circumstances which threatened to get out of hand. We do not only need to be sensitive to such issues but we also have to listen to what our employees have to say.

2. Effective communication

Effective communication with both customers and employees is a prerequisite for any successful organisation, but it is surprising how many companies focus almost exclusively on the former and neglect the latter. Bad communication within the organisation can be immensely destructive, leading to uncertainty, unrest and distress in the workforce. We need also to remember that imagination thrives on ambiguity and confused messages, mixed messages or half messages are just as bad as no message at all.

There are any number of ways of improving and maintaining effective communications within the organisation. Some management teams elect to have regular briefing sessions with senior management who can then go on to brief unit managers, who can in turn brief employees at shop-floor level. In other, smaller organisations, regular meetings with virtually the whole workforce is a realistic possibility. The method chosen will obviously vary from company to company, but the most important point is that improving the dynamics of communication is an extremely effective method of reducing organisational stress. There will always be some issues which cannot be discussed openly, whether for legal, commercial or other reasons, but provided this is explained clearly, employees are usually receptive and reasonable.

It is also important to bear in mind that communication does not simply mean sending letters and memoranda. A memorandum is no substitute for a one-to-one conversation which is infinitely more significant and revealing. Pieces of paper should be used to *support* communication, not as a substitute for it.

Many senior executives like to claim that they have an 'open door' policy, but you would be amazed at how few people go through it. The implication is that the onus is on the employee to go to the executive rather than the reverse. Good managers know the value of going out themselves to communicate directly with the workforce rather than leaving the initiative to the individual employee.

3. *Job description*

A clearly defined job description and line of responsibility are essential if ambiguity and potential conflict with others is to be avoided. Being unsure of one's responsibilities is a potent source of stress, and it leads to much unhappiness, anxiety and inefficiency. The job description should be written down as clearly as possible and the manager should ensure that the employee is fully conversant with his responsibilities and his reporting lines.

4. *Regular appraisal*

Performance evaluation is vital. People must be given the opportunity to grow and improve and develop their skills. This can only be achieved by regular assessment and feedback.

5. *Priorities*

In addition to an appropriate job description and regular appraisal, it is also important to establish a clear scale of priorities. Everything cannot be an A priority; some things have to be B or C priorities and some of the D priorities may not get done at all. Highly stressed employees are often on the receiving end of managers who try to make everything an A priority which then makes the whole concept meaningless.

Not only that; it means that the individual employee will always be the loser because no matter which one he chooses his boss can always claim that he should have chosen the other. It is rather like the story about the man whose wife buys him two ties for Christmas, a red one and a blue one. When he elects to wear the red one for dinner that evening, his wife says to him 'What's wrong with the blue one?'

6. Resources

Management must ensure that the necessary resources are available for the task to be accomplished. A great deal of worry is engendered by not giving adequate support, whether this be secretarial, financial, managerial or material. Agree beforehand what resources are required, and then ensure that they are made available.

7. Conflict management

Conflicts can have a variety of sources within an organisation and it is essential to have mechanisms by which they can be resolved. In an environment where you may have highly motivated, ambitious and competitive people working closely together, some degree of conflict is inevitable. Sometimes, a good hard straightforward argument is an excellent way of clearing the air and getting things 'on the table'. We should not shy away from this because, provided it is a contained reaction, it can be an immensely effective and therapeutic way of dealing with conflict.

However, this course of action is often not available to employees, and sometimes a grievance or a difference of opinion is more serious. It is of the utmost importance that each employee within the organisation should have available an alternative course of action, a mechanism by which an appeal can be made to a higher authority within the organisation. A clash of personalities between an employee and his immediate boss is very common, and it is important that the employee should not feel that if he suffers ill-treatment or victimisation he just has to accept it. An individual who has no appeal beyond his immediate superior becomes prey to intense feelings of frustration, anger and stress, making him a less effective employee and making the organisation less competitive.

Many organisations have a formal grievance procedure but others feel that less structured procedures are better. This tends to vary as a function of corporate culture, but the important thing is that it should exist and that employees should be well aware that they can take advantage of it if necessary and that they will not be victimised for so doing.

8. Setting objectives

It is imperative that there should be clearly defined objectives set within an overall strategic plan for the company. Moreover, these objectives must be clearly communicated to the workforce and each individual needs to be aware of the contribution he or she is required to make. A prerequisite for this, of course, is effective and decisive leadership, a point well made by John Harvey Jones: 'The business that is not being purposefully lead in a clear direction which is understood by its people is not going to survive, and all of history shows that that is the case.'

People often talk about 'management by objectives', but all too often these objectives are not clearly communicated to those individuals who are essential to the achievement of them. How often have we heard the complaint from employees that they want to help but they are not quite sure what it is they are supposed to do? Employees should not have to guess what their responsibilities are, or which set of objectives they are striving for. This only leads to ambiguity, uncertainty and confusion and can be intensely stressful. Given that most people *do* actually want to please, the responsibility of management is to give them the opportunity to do so. Make the objectives clear, unequivocal, with no room for misunderstanding.

9. Delegation

If an individual is to be given a new title with new responsibilities, it is essential that a commensurate degree of authority goes with them. Delegating does not mean 'I would like you to accept this new set of responsibilities but I will continue to make all the decisions'. That is again putting somebody in an intensely ambiguous and confused situation which can lead to uncertainty and very high levels of stress. You are, in effect, giving mixed messages. On the one hand you are implying trust and confidence in the individual's ability, but on the other you are suggesting that they are not quite up to it. If you think they are worthy, give them the authority and the responsibility, and then get out of the way; if you don't, then retain those responsibilities yourself. Alternatively, if you are prepared to delegate and give authority up to a certain *level*, then make it absolutely clear what that level is and set the limits of their authority and responsibility in clear and unequivocal terms. To be given responsibility without authority is one of the most intensely damaging situations in which any employee can find himself.

10. Selection

Extremely careful selection of individuals for promotion to more senior executive and managerial positions is of the greatest importance. A 'hit and miss' approach will not do. We are all aware of people who have been promoted to positions for which they have neither the ability nor experience. In some, incompetence is combined with an unfortunate tendency to abuse their position and authority, thus making the lives of their immediate subordinates an unremitting misery. Such individuals can have a catastrophic effect on morale and efficiency in a remarkably short period of time. It is obviously important that they should be identified and removed as soon as possible but, since prevention is better than cure, far better that they should never have been appointed in the first place.

Apart from using experienced personnel and structured interview procedures, some organisations nowadays are using more sophisticated means of psychological and personality assessment to help in making appointments.

11. *Environment*

Close attention to the working environment is essential if the workforce is to fulfil its potential. Apart from such obvious factors as heating, ventilation, lighting, a smoke-free environment and adequate rest areas, there is some evidence to suggest that natural light and plants enhance working performance. A poor working environment may be highly stressful and lead to increased absenteeism, impaired efficiency and reduced productivity.

12. *Job satisfaction*

We know from studies carried out by industrial and organisational psychologists that most people look for some kind of meaning in what they do, some opportunity for self-expression. In a real sense, therefore, the job becomes an extension of themselves. Many people quite enjoy mechanical, repetitive assembly-line types of occupation, but most people prefer variety and a real sense of progress in what they do. Those who want an exciting, challenging and rewarding job that allows them freedom and a chance to be creative, often find themselves forced into taking relatively dull and meaningless occupations. Carrying out a repetitive, mechanical and boring job can, contrary to popular belief, be intensely stressful. People who have never had to do this kind of work sometimes find it difficult to imagine how these sorts of tasks can be in any way regarded as stressful. The popular idea of stress being an exclusively executive complaint could not be further from the truth. Monotonous and repetitive jobs can be mind numbing and dehumanising. People look for meaning and some sense of identification with what they do, so having to act like a robot does nothing for their self-esteem. They start to feel they are not fit for anything more productive or meaningful. This can lead to a real sense of futility and become a potent source of stress and anxiety.

One of the features of assembly-line work is the individual's lack of control. His activity is dictated by the speed of the conveyor belt and the number of objects on it, over both of which factors he has very little control. He has become less of a human being and more part of a machine. Another feature of this kind of job is its isolation, its lack of a social dimension. One example will serve to illustrate these two features which are a source of stress and anxiety.

A very large company which specialises in canned meats has, in the course of a few years, changed beyond all recognition from the kind of factory where large numbers of people were employed to do a variety of jobs in preparing the meat for canning. This part of the industry is now highly mechanised and those who still work in the same factory now have very different jobs. One job, in particular, illustrates the lack of control and the isolation which are a feature of factory jobs today. One individual sits in front of an enormous conveyor belt with thousands of tins of meat passing in front of him. He is completely without company. Every so often, a red light appears on the console, in response to which the individual has to act immediately to find the fault on the assembly line (usually a can which has fallen off or jammed) and correct it as soon as possible. He then returns to his seat and waits for the next event to occur. This form of automation and mechanisation has allowed the company to maximise its profits while reducing the workforce by some 40 per cent.

Many companies in the United States have spent a considerable amount of time studying this problem and have come up with the concept of 'job enrichment'. The results indicate that even the most repetitive of jobs can be made more interesting, either by introducing a small element of variety or by extending the social interactions which take place in and around the job.

It is important to deveop a real understanding of what people are required to do at each level in the organisation, and to ensure that the task is made as tolerable and as rewarding as possible. Its sometimes worth asking what it would be like to have to do the job that you are asking somebody else to do. A few moments' reflection may show you how to enrich the job without in any way compromising the efficiency or the productivity of the individual concerned.

Individual strategies for stress control

There is no doubt that some people handle stress better than others, and we have all had experience of people who, no matter how complex and demanding their lives appear to be, always seem to be completely composed and in control. Various studies have revealed that these 'stress-resistant' individuals have a number of characteristics in common:

- A wide variety of interests and pursuits outside of their work which provide them with different challenges and satisfaction. For individuals who are less able to cope, work tends to be the sole source of gratification.
- Flexibility under stress. Stress-resistant people have a wide variety of ways of dealing with problems; when one way fails, it is not regarded as a defeat, but as an incentive to search for an alternative. They do not take any apparent failures personally. Problems exist only to be overcome and it is not a question of whether, but *when* they will overcome them.

- An ability to recognise and accept strengths and weaknesses. Stress-resistant individuals have a much more realistic view than others of their own and other people's abilities and limitations. If they fail to do something, they find it easier than others to accept it, learn from it and move on. They are not prone to self-reproach. They characteristically match their own weaknesses with the strength of other individuals.
- They take an active approach to problem solving which is extremely effective. They are able to establish priorities rapidly and then to focus intensely upon one issue at a time. More stress-prone individuals tend to put off major issues and approach problems passively. They are less able to set priorities and to concentrate their attention on each.

In developing an individual strategy for stress management, it is extremely important to remember that the psychological and physical parts of our make-up are not separate entities but are united in a single complex process. Thus our approach should emphasise the *whole* individual and not merely address his or her psychological or mental wellbeing.

Here are 12 basic principles which will help both you and the members of your workforce to become stress-resistant.

1. Avoid stressful situations

We often find ourselves asking the question 'Why did I get myself into this?' There are a great many situations which turn out to be intensely stressful which we could quite easily have avoided in the first place. There is no shortage of stressful situations – plenty for everyone. So accept only the essential ones and avoid the rest. *Before* you get into it, ask 'Do I want to get myself into this?'

2. Act

In many ways, one of the worst responses to a stressful or threatening situation is to do nothing. Even in the most difficult and trying of circumstances, something can be done, and this, in itself, can do much to reduce the stress.

We discussed earlier (p 219) the body's 'fight or flight' reaction to a dangerous situation. You will notice that this is an 'either/or' reaction, with no intermediate stage. Confronted by a 6ft 8in giant wielding a length of lead piping, your response is either to fight (bad choice), or to run as fast as you can (good choice). Trying to negotiate a settlement could be damaging. The point is a serious one. When someone is about to engage you in a furious argument, you have to stay and argue it out or simply take yourself away from the situation. Sometimes the latter approach is better because it immediately removes you from the source of stress and allows you to avoid saying things that you might later regret. On the other hand, a good straight hard-hitting argument may be the most appropriate response. Only you can make that judgement but the important point is that you should do *something*. The damage occurs when people try to adopt the intermediate position and do nothing. In these circumstances the tension and pressure continue to build without any mechanism for relief. Adrenalin and other stress hormones are released into the system to induce either a fight or a flight reaction. In the process of doing one or the other, these substances are used up and the stress response subsides. But if you do not do either, these substances can remain in the bloodstream and in the long term do a great deal of damage to the body.

So whether you are in an acutely stressful situation, or facing difficult and stressful problems, the advice is the same: 'Don't just sit there, *do* something!'

3. Organisation

Poor organisation lies behind a great deal of completely unnecessary and avoidable stress and anxiety. Review what resources you have at your disposal and see if you are utilising them to their full potential. It may be that the workload or the responsibility can quite easily be distributed more evenly among the group. Good organisation implies detailed forward planning and a clearly defined set of goals and objectives with an appropriate scale of priorities.

4. Effective time management

Once you have decided upon your priorities, do one thing at a time. Focus all of your effort on one task, and when you have achieved this, move on to the next. Do the important things sequentially and not concurrently.

Be realistic about your time allocation to various projects and the setting of deadlines both for yourself and for others.

5. Share your anxieties

The old adage that 'a trouble shared is a trouble halved' is an important principle in individual stress management. Sharing your anxieties and worries with those who are close to you is not a sign of weakness but is a necessary and important strategy. There are no brownie points awarded to those who carry their burdens in splendid isolation. Companies do not require martyrs.

6. Listen to others

When under stress we tend to lose our perspective of things. We can all recall intensely stressful events in our lives which, in retrospect, seem trivial and unimportant. We cannot imagine why they should have caused us so much stress at the time. The reason is that, in the middle of the situation, our perspective and judgements became seriously distorted. More often than not, we are unaware that this is happening to us, and it is only by listening to the advice of those around us that we are able to see how much things are getting out of hand. Friends and colleagues can help to restore a more healthy perspective, but only if we are prepared to listen to them.

7. Self-awareness

It is important to recognise signs of exhaustion and stress in yourself, and you should not ignore symptoms such as irritability, insomnia, depression and anxiety. Few people are able to recognise these symptoms at an early stage. Family, friends and colleagues may be far better judges of this than you are. If people around you are telling you that your behaviour is unusual and that your outbursts of temper are uncharacteristic, you must not simply dismiss them. These may well be early warning signs that you will ignore at your peril.

Best of all, develop a level of self-awareness which allows you to recognise the symptoms yourself and take appropriate action.

8. Relaxation

Having periods for relaxation and the development of new leisure interests is of the greatest importance. Many intensely work-orientated individuals suffer a real sense of guilt at taking time for relaxation and rest. They seem not to be able to understand that periods away from the working environment, doing something totally different, are essential periods of recuperation and revitalisation. Hobbies and interests outside the workplace are essential to the development of a balanced personality, and they are also extremely effective 'buffers' against the continuing stress of working life. Whether it is train sets or hang gliding the important thing is that you should have something in which you can feel absorbed.

More formal relaxation techniques such as yoga are based

upon controlled breathing and muscle stretching exercises. The best way to learn is to be taught by an experienced person, although there are plenty of tapes and self-help books on the market.

Another increasingly popular technique is that of meditation. The principle is very simple, but it has been shown in several studies to be a highly effective means of controlling stress and inducing relaxation. Again, the best way to learn is by direct instruction from an expert, but there are a number of excellent publications available which will allow the individual to develop the technique alone.

Athletes recognise that intense periods of training and activity need to be followed by periods of rest and recuperation, not only because it enhances physical performance and reduces the risk of injury, but also because it has a very positive psychological effect. The same principle works for all of us. Constant preoccupation with work is unhealthy and may actually reduce your effectiveness. Bertrand Russell put it well:

> one of the signs of impending madness is the belief that one's work is important. If I were a medical man I should prescribe a long holiday to anyone who believed that his work was important.

9. Regular exercise

This is an extremely important principle in effective stress management. Regular exercise not only promotes physical health and wellbeing, it also has an extremely powerful psychological effect.

10. Diet

When people are working under constant pressure, it is quite common for eating habits to deteriorate. The tendency is to resort to eating 'fast food' or 'junk food' which is always high in calories, fat and sugar but actually has little nutritional value. This can lead to deficiencies of important vitamins and trace elements, factors which may themselves adversely influence mood.

One of the other common dietary consequences of prolonged stress, is excessive coffee consumption. The average cup of brewed coffee contains about 140 milligrams of caffeine, so if you drink ten cups of coffee a day your total daily intake is around 1400 milligrams. Bear in mind that tea also contains significant amounts of caffeine, too much of which may cause anxiety, restlessness, tremors, insomnia, headaches and also produce an increase in heart rate, blood pressure and blood fats. While it is true that caffeine heightens physical performance, its effect on mood and intellectual performance is, contrary to popular belief, much less predictable. If you are a heavy coffee or tea drinker, you would be well advised to reduce your consumption.

11. *Alcohol and cigarette smoking*

Prolonged stress may also lead to an increase in cigarette and alcohol consumption. Alcoholism is one of our major contemporary health problems and is often precipitated by a desire to counteract the effects of stress. Drinking becomes an escape mechanism, a way of avoiding having to face stressful situations. It is remarkably easy to slip into regular and excessive alcohol consumption or to start smoking more cigarettes, which is why you should be extremely vigilant.

The need to find temporary respite from stressful life events may also lead to an increased dependence on prescribed drugs like tranquillisers or to a temptation to use non-prescribed drugs. Tranquillisers have their uses in certain specific clinical situations, but their long-term use cannot be recommended because they have significant addictive qualities and also a well-described withdrawal syndrome. In any event, these drugs do not solve problems, they merely postpone them, and the more they are postponed, the more difficult they become to deal with. It therefore becomes a dangerous vicious circle.

12. Adequate sleep

People vary enormously in the amount of sleep they require. Some feel refreshed after as little as 5 or 6 hours, while many others may need a minimum of 9 or 10. The quality of sleep, and not merely the quantity, is also an important factor and there needs to be a balance between dreaming sleep (called REM sleep because it is accompanied by rapid eye movements) and deep (non-REM) sleep.

We all have periods when we sleep badly. It may be due to a change in circumstances, such as a new baby, or examinations, or there may be no apparent reason. Usually these periods are shortlived, and we are quite able to cope with them. On occasions, however, stress may induce quite long periods of insomnia, and this may have quite significant long-term consequences including exhaustion. The more exhausted you become, the more aroused you become, and although you may feel that you could sleep for a year, the moment your head touches the pillow your mind is off again racing away with new ideas and new thoughts. Sleeplessness is due primarily to a heightened state of arousal and awareness which is sometimes extremely difficult to switch off.

Sleeping tablets will occasionally help you to get through a short-term difficult period, but they are not a recommended long-term solution because they do not always induce a natural form of sleep.

One of the most effective ways to deal with the problem is to learn formal relaxation techniques. There are many classes, tapes and books available, and although they require some discipline and confidence, they really do work if you are prepared to give them a try. Coffee, red wine and cheese can all promote sleeplessness and should be avoided; and remember that regular exercise also helps to induce relaxation and thus promote natural sleeping patterns.

Some people have no problem in actually getting off to sleep, but, if for any reason, they wake during the night, thoughts and ideas start to race around once again, making it very difficult to get back to sleep. The worst thing that you can do in these circumstances is to worry about the fact that you cannot get back to sleep. Anxiety is the worst enemy of sleep because it produces a heightened state of arousal. It is important to try and discipline yourself not to start thinking. This is obviously more difficult than it sounds, but over a period of time you can control your mind in such a way that you can empty it of the ideas and the worries which are preventing you from sleeping normally. If you find that you simply cannot get back to sleep, the best thing to do is to get up and make yourself a warm drink and perhaps listen to some music or read for a short while. This often relaxes you enough to return to bed for another few hours' rest.

We have looked at stress management principles for the organisation or company and for the individual. In practice, the strategies are closely interrelated, and it is obvious that the strategy most likely to succeed is one which takes into account both sets of principles. (See Figure 13.1.)

Implementation of stress management strategies

Organisational/corporate stress management

The successful implementation of stress management strategies at the organisational level ultimately depends upon the commitment and the resolve of senior management. First of all there has to be an awareness that the problem exists, and this usually requires one senior executive to bring it to the attention of his colleagues. Once a sufficient level of awareness has been achieved, there are a number of ways

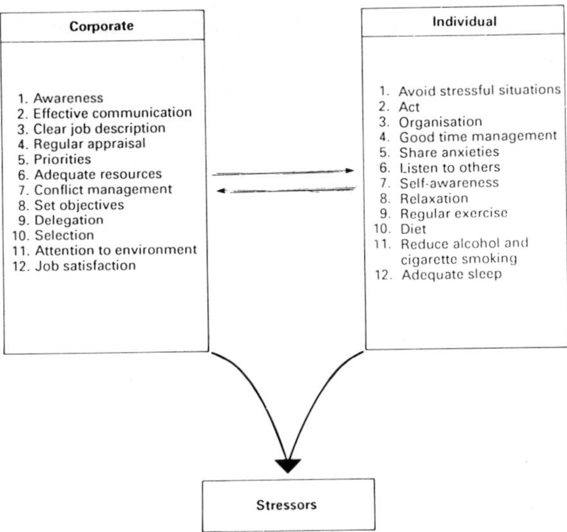

Fig. 13.1 *Effective stress management*

of proceeding. One useful suggestion is to develop a stress management working party, which will address at least three fundamental questions based upon the principles discussed earlier.

1. As senior managers, are we doing everything possible to reduce to a minimum those sources of stress at an organisational/corporate level, which can exert a negative impact upon the performance and wellbeing of our employees (including ourselves)?

2. What are we doing to create a greater level of awareness among the workforce as to what they as individuals can do to manage stress more efficiently in their own lives?

3. Are we, as individuals, setting a good example to our employees and managing stress effectively ourselves?

Individual stress management

Of course the decision to do something at an individual level rests ultimately with the individual, and knowledge itself does not necessarily lead to changes in behaviour. But knowledge is a prerequisite for change, and there is a great deal that management can do to increase the level of knowledge among the workforce as to what they as individuals can do in terms of managing stress more effectively within their own lives.

The ways in which an organisation does this depends upon the degree of commitment and the resources available.

Perhaps the simplest way of increasing awareness among the workforce is to make available to them literature outlining the basic principles for individual action which we discussed earlier. These can be prepared in-house, or by an outside agent. Either way, it is not an expensive item to produce. This not only increases awareness of the issues involved, but also emphasises the fact that the management of the company is aware that stress problems may exist and is prepared to do something about it.

A second level of approach is to provide seminars, lectures and discussion groups, supported by appropriate literature. This requires the services of an expert in the field, and you should be able to obtain some assistance with this from your local Health Education Authority or your local general practice. The individual can be retained on a sessional basis depending upon the size of the programme you wish to introduce, and can be extremely helpful in the design and preparation of literature, as well as in setting up discussion groups, seminars and lectures.

The most sophisticated and expensive approach is to establish a more formal and structured stress management programme by employing the services of an outside agency specialising in this area of medicine. Any such organisation should be under the direction of a psychologist with expertise in industrial or organisational psychology. Programmes normally involve three major elements:

- Initial screening
- Interpretation and identification of high risk individuals
- Individual counselling by a properly trained stress councillor or, in more serious cases, by a trained psychologist.

The initial screening process is quite often computer-based and may use one of the many widely available stress assessment inventories or questionnaires which also address personality and attempt to identify depressive illness. Examples are the CPI (California Psychological Inventory), EPI (Eisenck Personality Inventory), GHQ (General Health Questionnaire) and the HAD (Health Assessment Directory). These are highly complex questionnaires which require scoring and inter-pretation by an appropriately qualified individual. If you do decide to have this form of screening carried out, then it is extremely important that you should understand exactly what the questionnaire purports to measure and whether it has been internationally validated.

One useful approach adopted by some organisations is to test a small cross-section of the workforce, on the basis of which it is possible to estimate the levels of stress-related problems likely to be present in the workforce as a whole. This allows an organisation to make a more rational decision about how best to employ whatever resources it has available.

Another option worth considering is to employ the services of a stress counsellor for, say, one afternoon a week, and make this session available to any employees who wish to use it. This is along the lines of some Employee Assistance Programmes which are becoming increasingly popular in the USA, although it is difficult to know whether they will develop as rapidly in this country. It is often dangerous to extrapolate from the behaviour of individuals in one country to those in another. It may be that Americans are much more used to discussing stress-related problems than we are in the

UK. But the only real way of knowing whether such an initiative would be welcome and well used, is to try it and see. After all, the employment of a stress counsellor for one afternoon a week for a pilot project of say four to six weeks, is not expensive. If the programme does turn out to be popular, it can be an extremely effective way of dealing with problems at a very early stage, before they begin to have an adverse effect upon the individual and the organisation.

Summary

The implementation of an effective overall strategy for stress management depends upon a high degree of commitment on the part of the organisation and the individual, both of which have clearly defined responsibilities. Senior executives can do much to reduce corporate/organisational stress by close attention to good management practices. In addition, the organisation itself can do a lot to promote a greater awareness and a sense of responsibility on the part of individual employees to manage stress more effectively within their own lives. There are several ways of achieving this, from the simple provision of appropriate literature, to the setting up of formal and comprehensive stress management programmes with the help of an outside agency. Other models, like Employee Assistance Programmes may also prove to be effective in the future.

CHAPTER 14

AIDS (ACQUIRED IMMUNE DEFICIENCY SYNDROME)

'We stand nakedly in front of a pandemic as mortal as any pandemic there has ever been'.
Dr H Mahler, Director General, WHO, 1987

The acquired immune deficiency syndrome, AIDS, was first reported in 1981, and has since developed into a pandemic which now affects all continents and every country in the world. In Chapter 3 we saw that estimates of the progress of this disease, and even of the number of individuals currently infected, vary. But one thing is certain: AIDS will profoundly and irreversibly alter the social, psychological and moral fabric of our society.

The virus which causes AIDS was first identified at the Institut Pasteur in Paris, and is called the human immunodeficiency virus (HIV). HIV attacks a specialised group of cells in the body called T4 cells. In the normal, healthy human being, infections are kept at bay by an immune 'surveillance' system which is co-ordinated by the T4 cells. (See Figure 14.1.) When these T4 cells are attacked by the HIV, the body's immune system collapses, leaving the individual susceptible to a wide variety of infections especially pneumonia, and certain forms of cancer. (See Figure 14.2.)

The AIDS virus can also directly attack brain cells and, as a result, produce almost any kind of neurological syndrome and a variety of psychiatric illnesses marked by personality change, memory loss, intellectual impairment and dementia. Doctors now estimate that about 60 per cent of patients with AIDS will develop dementia at some time during their illness.

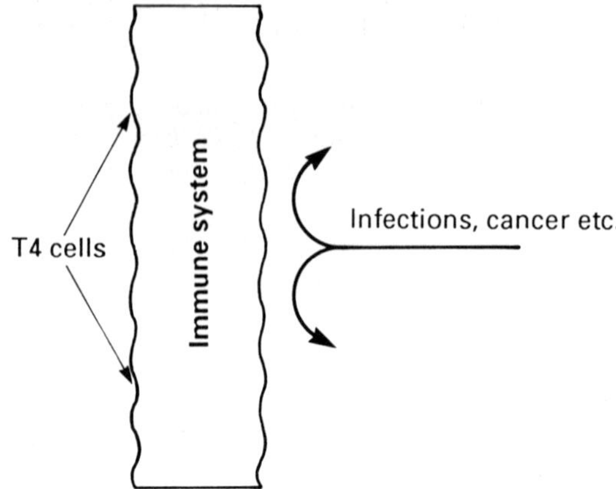

Fig. 14.1 *The immune 'surveillance' system of T4 cells in healthy human beings*

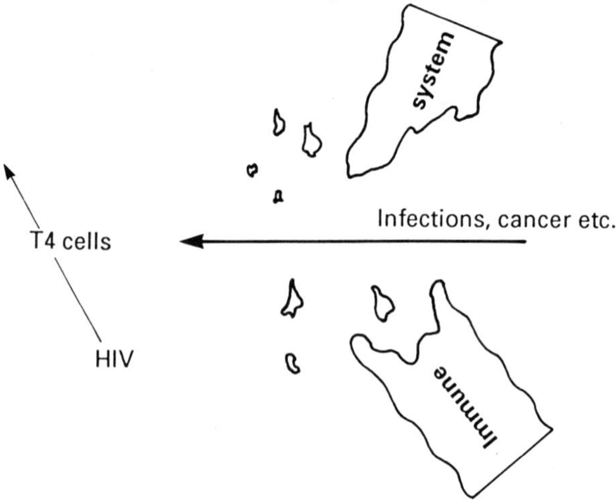

Fig. 14.2 *The immune system collapses when T4 cells are attacked by the AIDS virus*

Transmission is only through infected blood or blood products and through sexual contact involving exchange of body fluids. Women who have the virus can pass it on to their baby during pregnancy (through the placenta), at birth or through their breast milk. The risk of transmission to the baby seems to depend on the stage which the disease has reached in the mother. Infected women who have no symptoms probably have the lowest risk of infecting their child, probably in the order of 20 per cent.

The AIDS virus has been found in blood, semen, saliva, tears, breast milk, cervical secretions, cerebrospinal fluid and amniotic fluid.

Who are those most at risk?

The 'high risk' groups are now familiar to most of us and are as follows:

- Male homosexuals and bisexuals
- Men or women who are prostitutes
- Intravenous drug users who share needles and syringes contaminated with blood. This is one of the major sources of transmission in the heterosexual population.
- Haemophiliacs, who receive blood or blood products (mainly Factor 8). In 1982 it was shown that the transfusion of infected blood and certain blood products was responsible for infecting a number of individuals, including haemophiliacs, with the AIDS virus. Since October 1985, all blood donations have been routinely screened for the AIDS virus and blood products, such as Factor 8, are heat-treated.
- Individuals with Central African sexual contacts. AIDS is especially common in Zaire and Uganda, and is spread primarily by heterosexual contact. Anyone who has visited or lived in these areas and had sexual contacts belongs to a high risk group.
- Anyone with 'many' sexual partners. Difficult to define,

but it is certain that the more sexual partners you have the greater the risk.

- People with sexual partners in any of the above groups.

What is the natural history of the disease?

After the initial infection has occurred, there may be a brief flu-like illness, after which the individual enters the asymptomatic period, apparently perfectly well and leading a normal life. There is enormous variation in the length of this period, and it can be many years before symptoms of AIDS develop. It is also possible that some individuals may never actually develop AIDS and will simply remain in the completely symptom-free carrier state.

On average, about half of all HIV positive patients will develop some clinical signs or symptoms of AIDS within five to six years of infection. Although insufficient time has elapsed to be absolutely certain, many experts believe that all patients infected with HIV will eventually develop AIDS. When the syndrome does develop, it is invariably fatal; 80 per cent die within two years and 95 per cent within five years. The natural history of the disease is summarised in Figure 14.3.

Testing for infection with the AIDS virus

The commonly used test, known as the ELISA method, does not detect the virus itself but the antibodies produced by the body in response to infection with the virus. There are other, more complex types of test available, but they take longer to perform, are more expensive and cannot currently be used for mass screening.

The ELISA test is highly sensitive but not very specific, and this means that although the number of false negatives is very small, the number of false positives is high. All positive tests done by the ELISA method must therefore be confirmed by a more specific test designed to identify the virus itself. No patient should ever be informed that he or she is AIDS positive, unless and until this confirmatory test has been carried out. In some cases, the interpretation of the test result remains in doubt, and it may be necessary to retest the person a few months later.

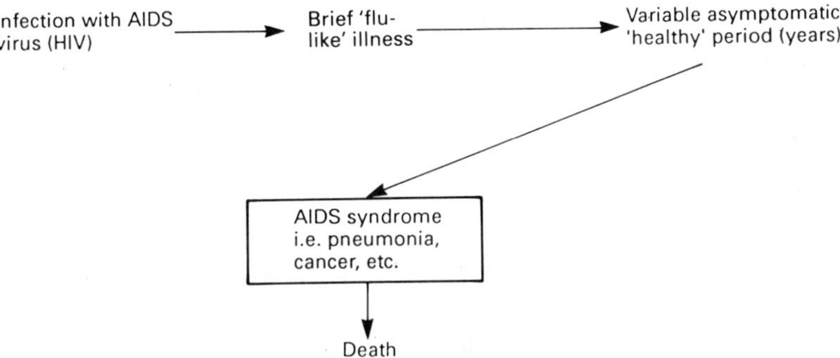

Fig. 14.3 *The natural history of AIDS*

Antibody to the AIDS virus usually takes about three months to appear in the bloodstream after the time of the initial infection. This means that an individual may be an HIV carrier but test antibody negative. Moreover, some individuals take more than three months to develop the antibody and there are even some individuals who will be long-term HIV carriers and *never* develop the antibody.

Interpretation of the HIV antibody test

Positive test: A confirmed positive test means that a person has antibodies to HIV in his or her blood and is therefore carrying the AIDS virus. In other words, the person is infected and also potentially infectious. It does not mean that the person has AIDS but merely that he or she is an HIV carrier. Neither does it give any information as to the future likelihood of developing AIDS syndrome or as to when or by what means the original infection occurred. In summary, the only implication of a confirmed positive antibody test is infection by the HIV.

Negative test: A negative result means a person does not have detectable amounts of antibodies in their blood. This may either be because they are not infected with HIV, or that they have been infected but have not yet developed the antibodies. It follows that a negative HIV antibody test does not *necessarily* indicate the absence of the AIDS virus.

Testing for AIDS and consent

To carry out any procedure on a patient without consent, constitutes an assault. The giving of a blood sample carries with it the implied consent for certain tests to be carried out in the interests of the patient's health. However, a test for HIV antibodies would not be considered routine because of the serious medical and social implications associated with it. It follows that if explicit consent is not obtained before testing for the AIDS virus, the doctor may be legally liable.

The need for counselling

Aside from the question of consent, all patients should be counselled in detail as to the potential consequences of a positive antibody test for HIV. Some of the more important points to stress are as follows;

- A negative test is no guarantee that the AIDS virus is not present, and nor does it prevent future infection.
- A positive test does not mean that one has AIDS, but it does mean that one is infected and infectious to others. A positive result can lead to problems with housing, social life, employment and insurance.

- Even a negative test result can lead to problems with insurance policies and mortgages if the insurers think that an individual is at risk.
- A positive test means that there is a greatly increased risk of developing a chronic, debilitating and ultimately life threatening disease at any time within the next ten years.
- There is no cure for AIDS and there is no proven treatment for a symptom-free person who is carrying the AIDS virus. There is a significant risk that the unborn baby of an infected woman may also become infected.

Confidentiality

It is self-evident that the issue of confidentiality is of fundamental importance when testing for AIDS. All medical information on an individual should be regarded as confidential but this is particularly so in relation to the HIV antibody status. This is because discrimination may result, not only if an individual is known to be HIV positive, but even if he or she is known to have been tested or has merely received counselling.

Implications for employers

AIDS raises a myriad questions for employers. The three most important, which we will now consider, are as follows:

1. How should we manage those employees who turn out to be carriers of the AIDS virus or who develop full blown AIDS?
2. How should we address employee attitudes to AIDS victims?
3. What about recruitment and the question of HIV screening?

The HIV positive individual

Under normal, everyday working conditions, the AIDS sufferer or HIV carrier is unlikely to represent a threat to his colleagues.

From a legal point of view, ill-health is a fair reason for dismissal, but usually only when the illness renders an individual incapable of performing his normal duties, or when it renders him a significant threat or risk to his fellow employees or the customers of the company. In the case of the HIV positive individual, neither condition would normally appear to apply, because most HIV positive individuals are in good health, and they do not normally pose any risk to their colleagues. Dismissal purely because a person had developed the antibody would probably not, therefore, be regarded as fair in law.

If an individual develops AIDS syndrome, dismissal would still only be fair if he were incapable of performing his normal duties, or if his condition directly or indirectly caused significant risk to his colleagues, customers of the company or himself. In the vast majority of cases, the criteria used do not differ from other cases of serious illness.

It can be seen that in general terms the rights and obligations of employees and employers are unchanged and the mere existence of AIDS generally makes no difference to employment practice.

There have, however, been several occasions on which the Employment Appeal Tribunal has judged dismissal to be fair when an employee was *believed* by his employer to pose a risk to his colleagues, even though no scientific evidence was offered to support such a view. A particularly disturbing example was the case of a cinema projectionist, a known homosexual, who was dismissed because of the fear among some of his colleagues that he posed a serious AIDS risk. His dismissal was subsequently upheld by an industrial tribunal despite the fact that he was not actually an AIDS sufferer nor was he carrying the AIDS virus.

Cases of this kind are exceptions to the rule but they do suggest that the approach of the tribunals in this area is a little 'loose' to say the least. Perhaps a lack of knowledge of AIDS and the HIV virus at that time may have been a factor. It certainly supports the view that homosexuals have, in reality, very little protection under the law as it stands. Nevertheless, it can be summarised that:

- If an employee is HIV positive this would not, in itself, normally be grounds for dismissal.
- If the HIV positive individual develops the syndrome, then the question of dismissal can only be fairly considered when the illness has rendered the individual incapable of performing his normal duties, or if his condition directly or indirectly places him, his colleagues or the company's customers at risk.

Employee attitudes

Much of the fear about AIDS is irrational and based upon wrong information and rumour. Situations arise in which people object to working with someone with AIDS or with someone who is an HIV carrier, or even with someone whom they believe *may* be carrying the AIDS virus. The best way to deal with this is by education, discussion and reasurrance. Providing factual information, especially about methods of transmission, can do much to allay fears. It is better to introduce educational programmes *before* the problem occurs, rather than as a reaction to a specific incident.

Some companies and organisations argue that the obligation to educate individuals about AIDS rests firmly with the government and not with the employer. This is a shortsighted and misplaced view which completely underestimates the scale of the problem. Responsibility to educate rests with all of us and employers are particularly well placed to deliver a consistent and factual 'AIDS avoidance' programme to their employees. There is no question that the most important task is to educuate the workforce.

If an employer fails to deal reasonably with irrational fears, and succumbs to pressure or threats to the extent of dismissing a particular employee on the grounds that he may be an AIDS risk, it is likely to be regarded in law as unfair dismissal. Moreover, suspending such individuals may well serve only to reinforce the groundless fears of their colleagues.

Recruitment and screening

In general terms, employers have a free hand concerning who they wish to employ, save for the restrictions consequent upon the Sex Discrimination Act 1975 and the Race Relations Act 1976 and particular requirements regarding professional qualifications. Ill-health, actual or potential, is a reason in the UK for which an employer has the right to refuse to employ someone. There is no law which prohibits this. So someone who is known to be HIV positive or who has AIDS, may legitimately be refused employment. Of course, in the great majority of cases, the employer does not know – a virus carrier may have no symptoms at all. This raises the difficult question of pre-employment screening for AIDS.

At first sight, the logic of pre-employment AIDS screening would appear to be sound and reasonable. It offers the employer the opportunity to avoid taking on someone who may become ill, or who may cause disruption in the workplace if found, or suspected to be, an HIV carrier. Pre-employment testing, it is argued, could well save the company significant amounts of money in the future.

But it is not a simple matter. I believe that there are a number of problems with this kind of reasoning. First, such testing is always bound to be coercive. However professional and unbiased pre-test counselling appears to be, the implicit threat is 'If you do not have the test, you will not be employed'. Such coercion is ethically questionable.

Second, as we have seen previously, screening does not

identify all those who are infected and it cannot identify any of those who will become infected *after* employment. Now, the fact that you cannot guarantee the value of a negative test, does not in itself negate the value of a positive test. But the potential number of positive tests at the present time, is so small that the value of testing on such a large scale is very questionable. We do not have an accurate figure for prevalence rates in the population as a whole, but there are nevertheless some indicators. For example, since 1986, records have been kept of the total number of first-time blood donors who have been tested for HIV antibody. Of the 800,000 tested, there have been 29 confirmed positives, that is one in 28,000 or 0.0036 per cent. If this reflected the prevalence rates in the population as a whole, the financial argument for pre-employment AIDS screening would be difficult to support.

The third point is that it is extremely important to set the AIDS problem in context. While it cannot be denied that there is a very high probability that the HIV positive individual will suffer premature disability or death, this must be compared with potential for premature morbidity and mortality from a variety of other diseases. For example, it seems likely that there will have been 10,000 and 30,000 AIDS cases diagnosed by the end of 1992 in England and Wales. But coronary heart disease alone kills 160,000 people every year in this country, and disables at least the same number. Add to this the human and economic consequences of such conditions as stroke, respiratory disease, lung cancer, drug addiction and alcohol abuse and the AIDS problem is seen in a different perspective.

Many companies have chosen to treat AIDS like any other serious or fatal disease. Control Data Corp in the USA is typical of these companies. It conducted a study which showed that, even on the most pessimistic estimates, just over 100 of its 34,000 employees would die of AIDS during the next five years – fewer than will die of other major killers such as cancer and heart disease. In view of this, the company does not conduct any routine screening measures but has simply decided to include AIDS sufferers within the overall company policy for individuals with serious or life threatening disease. Many British companies also take this view and this seems reasonable, especially as the number of AIDS sufferers in this country at least at the present time, is substantially less than in the USA.

It has been argued that in certain occupations, particularly where public safety is involved, such as train drivers, airline pilots and so on, HIV screening is required because of the possibility of brain damage leading to impaired judgement. But there is little evidence to support such a contention. The World Health Organisation in 1988 looked at links between mental deterioration and HIV, and concluded that there was no evidence of clinically significant mental abnormalities in people infected with HIV who were otherwise free of symptoms. Significant intellectual impairment *follows* but does not precede other clear indications of AIDS syndrome which would preclude the individual from carrying out normal duties. The WHO, while recognising the importance of identifying workers whose mental capacity is impaired, also stressed that mental deterioration due to factors *other* than the AIDS virus, is much more common.

In summary, it is doubtful whether pre-employment AIDS screening could be justified at present, simply because there is no good evidence that either the employer or the employee would benefit from it. Whatever resources are available within the organisation could be used far more profitably in establishing an AIDS educational programme, designed to prevent individuals contracting the disease in the first place.

Preventing AIDS

Since 1981, when the AIDS syndrome was first described, there has been an unprecedented effort on the part of the international scientific community to find answers to the formidable problems which the phenomenon poses.

It seems reasonable to suppose that a vaccine will, eventually, be produced, but most experts do not believe that a vaccine will be generally available for at least another five to ten years. Moreover, although progress has been made in the development of drugs for the treatment of AIDS, it is extremely unlikely that they will ever constitute a cure. The most that we can expect to achieve from drug therapy is that the disease will be slowed down and life expectancy prolonged.

While the race to find a vaccine and effective treatments for AIDS continues, the most effective means of controlling this epidemic at the present time is through education. AIDS is preventable. The challenge is to get across the message that it is what *individuals* do and how they behave that matters.

The basic principles of prevention are familiar, but they bear repeating:

- In sexual relationships other than with your regular partner, practise safe sex and use a condom.
- Avoid anal or oral sex.
- Avoid sex with individuals from high risk groups such as prostitutes and drug addicts.
- Avoid any activity which may tend to break the skin or draw blood.
- Avoid using other people's razors, toothbrushes and toiletries.

There is now an enormous amount of information available on the subject of AIDS prevention, including a number of tapes and videos. It is important to ensure that all members of the workforce are given access to this information and that it should be reinforced at regular intervals. Discussion groups, seminars and lectures are useful ways of promoting a greater level of awareness among the workforce.

AIDS and first aid

The risk of catching AIDS from the application of normal first aid procedures is extremely remote, but first aiders may well need reassurance. Employers are required by law to provide first aid, and this may involve the provision of trained first aiders or the appointment of some other person to take charge of the situation where an employee is ill or injured at work. First aiders obviously need to be given clear and appropriate information about any possible risks relating to the service they give, and it is the employer's legal responsibility to provide this.

It is basically a question of reinforcing the hygiene precautions which are already an integral part of standard first aid procedures. Some additional items of equipment are also useful, such as disposable plastic gloves, protective plastic aprons, disposable plastic waste bags and special mouthpieces for resuscitation. The CBI and other organisations have published detailed guidelines in relation to AIDS and first aid procedures, and these are available to all employers.

AIDS and travel abroad

The employer also has a responsibility to ensure, so far as possible, the health and safety of employees when they travel abroad. This means that they must be provided with information about the potential risk from the AIDS virus, and be given specific instructions as to how to obtain medical help in the case of illness or accident in the countries which they intend to visit.

Although the World Health Organisation and the British Medical Association advise against screening, some countries, such as Saudi Arabia and Iraq, require it for foreign visitors. The employer therefore needs to check whether a certificate is required for a particular country and make the necessary arrangements. It is important to bear in mind that even in routine screening situations such as this, appropriate counselling is mandatory before the test for the HIV antibody is carried out. Once again, the CBI, the Health Education Authority and the DHSS have prepared detailed guidelines concerning the issue of AIDS in relation to business travel overseas.

Formulating a company AIDS policy

One of the problems with the AIDS epidemic is that the natural history of the disease, particularly the long incubation period, may tend to breed a certain amount of complacency. A continuing educational programme is therefore extremely important, and regular reinforcement of the health education message to the workforce should be an integral part of any company policy on AIDS. This is not to suggest that the subject needs to be talked about on a daily basis: too much of the same kind of information can actually have a negative effect. Once the level of awareness among the workforce is high, it simply needs regular 'topping up'.

It is useful for a company to develop a specific written policy on AIDS, not only because it obviously provides useful guidelines, but also because it encourages managers within the company to address many of the difficult issues which the AIDS problem poses. As with alcohol, drugs and smoking policies, there is no single blueprint which can be applied to all companies, since they vary enormously in terms of their specific requirements, attitudes and resources. However, there are five areas which should be addressed by all organisations and these are listed below. It is an advantage to have a small working group to define the policy, but the overall responsibility should be given to one named individual.

1. *Education:* The working group should decide what educational initiatives are required, plan an appropriate

programme and make the necessary arrangements for implementation. Help is available from many professional organisations.

2. *First aid:* In the light of the published guidelines, the working group should ensure that first aiders are provided with any additional information or training which may be required and have ready access to any additional items of equipment which may be recommended.

3. *Overseas travel:* The working group should ensure that any employees travelling abroad, particularly to high risk areas, receive appropriate information and advice, and that the mechanism for providing medical support is in place *before* the employee travels. If an HIV antibody test is required for entry into a particular country, then the group should ensure that detailed expert counselling is given before the test is carried out.

4. *Pre-employment screening:* The company should state in clear terms what its policy is with regard to pre-employment screening for the AIDS virus.

5. *Managing HIV positive or AIDS victims:* The company should have a clearly defined method of dealing with the rare eventuality of an employee developing AIDS or with any employees who turn out to be HIV carriers.

POSTSCRIPT: BUILDING A HEALTHIER SOCIETY

'. . . . I think it is actually true that humanity will win eventually. However, I am afraid that at the same time the world will be a huge hospital and one will be the humane nurse of the other'.

Goethe

Throughout this book I have emphasised the fundamental point that what primarily determines our health is the choices that we make for ourselves. In this context, I believe that the role of the corporate organisation in encouraging healthier choices and in supporting healthier behaviour, is of considerable importance. There are compelling arguments for so doing, and there is also a moral and ethical imperative which cannot be ignored.

But, of course, this is only one side of the story. Individuals obviously have a measure of freedom and responsibility in the choices they make, but they cannot be held wholly responsible if they chose inappropriately. To argue that they should be is to ignore the fact that our choices are strongly influenced by others, like the advertising and marketing activities of many multi-billion pound organisations designed to ensure that our choices are good for business. The alcohol industry spends about £200 million a year on its promotional activities, and the tobacco industry well in excess of £100 million a year. The food industry also has massive advertising and marketing campaigns. When we set all of this against the £9.4 million allocated to the Health Education Authority, which is trying to encourage us to make healthier choices, it is not surprising that a great many of us continue to choose the other way.

Our educational background is also an important factor in determining the choices we make, and people are able to make much more rational decisions if they have the information made available to them in terms that they understand. It is often argued that if large sections of the population want to go on smoking, drinking and eating themselves into an early grave, they have only themselves to blame. No doubt this is true for many, but, in my experience, it is more often that they genuinely do not understand what the real issues are and what it is that they have to do. After all, parents do not *knowingly* damage the health of their children, and yet countless numbers of them are doing exactly that purely out of ignorance. We have to develop much more effective ways of communicating health issues to all sections of the population, including those who are less educationally advantaged. Knowledge, is not in itself sufficient to produce changes in behaviour, but it is a *prerequisite.*

Some choices are largely determined by the amount of money in our pockets, which is largely outside the control of the individual. It is quite clear that if a healthy diet means an expensive diet, then a large section of the population, including the unemployed, the elderly and the poor, will simply not have access to the right foods.

At the beginning of this book I stressed the need for a democracy of prevention. The danger is that we are creating a system in which the health education message is getting across to one section of the community who, by virtue of their social class, educational background and financial resources are able to take advantage of it, while the remainder continue as before. The evidence is that most of the current changes in exercise, smoking, diet and alcohol consumption are taking place predominantly in the middle class and well-educated sections of our society.

The responsibility of government in all this is clear. It is to ensure that *all* individuals have an opportunity to make healthier choices, irrespective of their educational background, social class or financial circumstances. Once that choice has been made available, it is up to the individual to act responsibly and take advantage of the opportunity.

The problem is, of course, that this would bring the government into conflict with various groups and organisations which have a vested interest in perpetuating the status quo, and have enormous economic power and political influence. In other words, what is good for the alcohol or food industry and for the national economy, is not necessarily in the best interests of public health. Improving the health of the population is desirable, but may also be bad for business and politically risky. That is why I believe that the whole issue of the national diet and the environment must be opened up to involve a much wider spectrum of opinion than it does currently. At the moment, it is precisely those who have the greatest financial interest in food production and food processing, including the farmers, who have the biggest say in shaping future policy. Not only do they set the agenda, but they largely determine the outcome. The Ministry of Agriculture, Fisheries and Food is supposed to represent the interests of producers, manufacturers and consumers, but the conflict of interests is obvious to anyone. It is rather like asking the fox to look after the welfare of the chickens. The public will not be reassured as to the safety of pesticides, hormones, additives, antioxidants, emulsifiers and colourings, etc if the reassurance is coming from an organisation that is also representing the interests of those who use these substances to boost production and profits.

There is an urgent need for an independent organisation *with teeth*, to represent the interests of the consumer with particular reference to food safety and public health. A ministry of food could be responsible for the establishment of a national food policy within whose framework the activities of the food and drink industry could be more closely monitored. More importantly, such an organisation could be the focus of a much closer dialogue between all interested parties, including farmers, food manufacturers, the Health Education Authority, the Department of Health, and the medical profession. One of its first tasks should be to define a national dietary policy designed to promote healthier eating habits among all sections of our society, including school children and the elderly.

Of equal importance in our efforts to build a healthier society, is the need to encourage a much wider debate about the future role of our current healthcare system and the NHS in particular. At the moment the major participants are the government, GPs, consultants, nurses and other healthcare professionals, all of whom have a vested interest in a particular outcome. The consumer – the patient – sits on the sidelines, awaiting his fate. The fact is that the health debate must be taken out of the hospitals, out of the surgeries and clinics and into society as a whole so that the consumers themselves have the fullest possible opportunity to participate. After all, the future health of the nation is far too important an issue for it to be left merely to doctors, politicians and other health professionals. Indeed, if we had managed to achieve a deeper and more fundamental debate about the system itself, then we might have come up with a much more radical and effective solution than that which is on offer by the government for the 1990s. No real alternative to our current sickness service is suggested, only alternative means of delivering it. This is hardly surprising since all those participants in the current debate have one fundamental assumption in common: that more medicine, more medical technology, more nurses, doctors and hospitals invariably means a better standard of health for the nation as a whole. It is simply a question of finding the right financial and administrative structure to make it work.

The concern that I and many others have is that in our constant struggle to obtain *more* medical care, we are ignoring the more important question: What made us ill in the first place and how much of this sickness could have been avoided? Instead of arguing about how we can deliver a more cost effective *sickness* service, we should be discussing how we can build a more comprehensive *healthcare* system that will include the promotion of health and the prevention of disease as one of its most important objectives. This was precisely the vision contained in the Beveridge Report of 1942.

A company-based programme of preventive medicine could provide the first line of care in a new system. Not only will this help to promote and maintain the health of the workforce, but by extending these initiatives into the local community, it is possible to bring about better standards of health for many others. We know that this is medically effective and there is growing evidence that it makes good economic sense too. Just as important is the fact that it will help us to direct healthcare away from a system which is dominated by the treatment of disease, towards a more balanced approach which has equal regard to the need for prevention.

This first line of care must be supported by, and integrated with, the conventional medical services for the sick and injured, provided by general practitioners, hospitals and other institutions both in the NHS and in the private sector. The GP in particular, is crucial to all of this, since he or she can not only provide treatment and support for the sick, but can also work with companies and organisations in support of preventive initiatives in the workplace and the community. This will eventually result in a reduced burden of disease in the population as a whole, which will relieve much of the pressure on the NHS and allow it to function more effectively at what it does best, that is treating sick and injured people.

Last, but by no means least, these endeavours must be underwritten by a genuine commitment on the part of government, the food and alcohol industry, the pharmaceutical industry, the Ministry of Agriculture, Fisheries and Food, the Department of Health and the medical profession itself, to build a better healthcare system for everyone. While recognising that this will require an enormous change in attitude on the part of many of those concerned, I believe that such a change is a prerequisite for a new beginning.

As Beveridge says in the opening sentences of his report: 'A revolutionary moment in the world's history is a time for revolutions, not for patching.'

Appendix

USEFUL ADDRESSES

Sources of various publications and reports

British Medical Association
Tavistock Square
London WC1H 9JR
Tel: 01-387 4499

Cancer Research Campaign
2 Carlton House Terrace
London SW1Y 5AR
Tel: 01-930 8972

Centre for Health Economics
University of York
Heslington
York YO1 5DD
Tel: 0904 430000

Centre for Policy Studies
8 Wilfred Street
London SW1E 6PL
Tel: 01-828 1176

Confederation of British Industry
Centre Point
103 New Oxford Street
London WC1A 1DU
Tel: 01-379 7400

Health Promotion Research Trust
Assets House
17 Elverton Street
London SW1P 2QG
Tel: 01-834 1427

HMSO Publications Centre
PO Box 276
London SW8 5DT
Tel: 01-873 9090/0011

Incomes Data Services Ltd
193 St John Street
London EC1V 4LS
Tel: 01-250 3434

Institute of Economic Affairs
Health Unit
2 Lord North Street
London SW1P 3LB
Tel: 01-779 3745

Office of Disease Prevention and Health Promotion
Room 2132, Switzer Building
330 C Street, SW
Washington DC 2021
USA
Tel: 202-245 7611

Office of Health Economics
12 Whitehall
London SW1A 2DY
Tel: 01-930 9203

Office of Population Censuses and Surveys
St Catherine's House
10 Kingsway
London WC2B 6JP
Tel: 01-242 0262

Royal College of Physicians
11 St Andrew's Place
Regent's Park
London NW1 4LE
Tel: 01-935 1174

Royal Society of Medicine Services Ltd
1 Wimpole Street
London W1M 8AE
Tel: 01-408 2119

The Bureau of National Affairs Inc
1231 25th Street, NW
Washington DC 20037
USA
Tel: 800-452 7773/202-452 4323

Advice on health education in the workplace and health promotion materials

Action on Smoking and Health
5–11 Mortimer Street
London W1N 7RH
Tel: 01-637 9843

Alcohol Concern
305 Grays Inn Road
London WC1X 8QF
Tel: 01-833 3471

British Heart Foundation
102 Gloucester Place
London W1H 4DH
Tel: 01-935 0185

British Medical Association Foundation for AIDS
BMA House
Tavistock Square
London WC1H 9JP
Tel: 01-383 6345

Chest, Heart & Stroke Association
Tavistock House North
Tavistock Square
London WC1H 9JE
Tel: 01-387 3012

Coronary Prevention Group
60 Great Ormond Street
London WC1N 3NR
Tel: 01-833 3687

Health Education Authority
Hamilton House
Mabledon Place
London WC1H 9TX
Tel: 01-631 0930

Institute of Alcohol Studies
Alliance House
12 Caxton Street
London SW1H 0QS
Tel: 01-222 5880/4001

Institute of Personnel Management
IPM House
35 Camp Road
London SW19 4UW
Tel: 01-946 9100

National AIDS Helpline
Tel: 0800 555777 for literature
0800 567123 for personal and confidential advice

Preventive medicine and health-screening

AMI Health Care Group plc
Corporate Health Services and Lifestyle Health-Screening
4 Cornwall Terrace
Regents Park
London NW1 4QP
Tel: 01-486 1266

Behavioural Science Consultants
C/O Prof HR Beech
Priory Hospital
Rappax Road
Hale, Atrincham
Cheshire WA15 0NX
Tel. 061-904 0050

BUPA Medical Centre (Men's Unit)
Webb House
210 Pentonville Road
London N1 9TA
Tel: 01-278 4651

BUPA Medical Centre (Women's Unit)
Battle Bridge House
300 Grays Inn Road
London WC1X 8DU
Tel: 01-837 6484

The London Heart Clinic
22 Upper Wimpole Street
London W1M 7TA
Tel: 01-486 8961

PPP Preventative Health Services
New Cavendish Medical Centre
99 New Cavendish Street
London W1M 7FQ
Tel: 01-637 8941

SELECTED BIBLIOGRAPHY AND FURTHER READING

Action on Smoking and Health 'Pipe and Cigar Smoking: Report of an Expert Group', *The Practitioner* 1973, **210** 645–52.

Anderson, P 'Excess Mortality Associated with Alcohol Consumption', *British Medical Journal* 1988, **297** 824–6.

Andreasson, S, Allebeck, P and Romelsjö, A 'Alcohol and Mortality Among Young Men: Longitudinal Study of Swedish Conscripts', *British Medical Journal* 1988 **296** 1021–5.

Anscombe, E and Geach, P T (1970) *Descartes' Philosophical Writings*, Nelson's University Paperbacks, The Open University, London.

Ashton, D and Davies , B (1986) *Why Exercise?*, Basil Blackwell Ltd, Oxford.

Ashton, H and Stepney, R (1982), *Smoking: Psychology and Pharmacology*, Tavistock Publications, London.

Babor, T F, Ritson E B and Hodgson R J 'Alcohol-related Problems in the Primary Health Care Setting: A Review of Early Intervention Strategies', *British Journal of Addiction* 1986, **81** 23–46.

Barefoot, J C, Dahlstrom, G and Williams R B 'Hostility, CHD Incidence and Total Mortality: A 25-year Follow-up of 255 Physicians', *Psychosomatic Medicine* 1983, **45** 59–63.

Beer, Stafford (1975) *Platform for Change*, John Wiley & Sons Ltd, Chichester.

Bellingham, R and Cohen, B (Ed) (1987) *The Corporate Wellness Sourcebook*, Human Resources Development Press Inc, Amherst, Massachusetts.

Betteridge, J D 'High Density Lipoprotein and Coronary Heart Disease', *British Medical Journal* 1989, **298** 974–5.

Beveridge, Sir William (1942) *Social Insurance and Allied Services*, HMSO, London.

Bly J L, Jones R C and Richardson J E 'Impact of Worksite Health Promotion on Health Care Costs and Utilization. Evaluation of Johnson and Johnson's 'Live for Life' Program', *Journal of the American Medical Association* 1986, **256** 3235–40.

Bureau of National Affairs Inc (1986) *Health Care Costs: Where's the Bottom Line?* Washington DC.

Burkitt, D (1984) *Don't Forget Fibre in Your Diet*, Martin Dunitz, London.

Cancer Research Campaign (1988) *Facts on Cancer*, London.

Capra, F (1983) *The Turning Point*, Fontana Paperbacks, London.

Committee on Medical Aspects of Food Policy (1984) *Diet and Cardiovascular Disease: Report of Advisory Panel on Diet in Relation to Cardiovascular Disease* DHSS Reports on Health and Social Subjects No 28, HMSO, London.

Confederation of British Industry (1987) *Absence From Work: A Survey of Non-Attendance and Sickness Absence, London.*

Confederation of British Industry (1989) Managing for Attendance: A Guide Prepared by the CBI Absence Steering Group, London.

Cook, J (1987) *Whose Health is it Anyway?*, New English Library, Hodder & Stoughton, London.

Corporate Health Care Management Company (1986) *Health Promotion Program Survey*, New York.

Cox, M, Shephard R J and Corfy P 'Influence of an Employee Fitness Program upon Fitness, Productivity and Absenteeism', *Ergonomics* 1981, **24** 795–806.

D'Arcy, Masius Benton & Bowles (1986) *The DMB & B Healthy Eating Study*, London.

Dawber, T R (1980) *The Framingham Study: The Epidemiology of Atherosclerotic Disease*, Harvard University Press, Cambridge, Massachusetts.

DHSS (1976) *Prevention and Health: Everybody's Business*, HMSO, London.

DHSS (1986) *Breast Cancer Screening: The Forrest Report*, HMSO, London.

Doll, R and Peto, R (1982) *The Causes of Cancer: Quantitative Estimates of Avoidable Risks of Cancer in the US Today*, Oxford University Press, Oxford.

Doyal, L and Pennell, I (1979) *The Political Economy of Health*, Pluto Press, London.

Dunbar, G C and Morgan, D D V 'The Changing Pattern of Alcohol Consumption in England and Wales 1978–85', *British Medical Journal* 1987, **295** 807–10.

Fentem, P H, Bassey, E J and Blecher, A (1979) *Exercise and Health: A Bibliography,* University of Nottingham Medical School, Nottingham.

Fleming, A F, Carballo, M, Fitzsimmons, D W, Bailey, M R and Mann, J (1988) *The Global Impact of Aids,* Alan R Liss Inc, New York.

Fraser, G B (1986) *Preventive Cardiology,* Oxford University Press, Oxford.

Friedman, M and Rosenman, R H 'Association of Specific Overt Behaviour Pattern with Blood and Cardiovascular Findings, *Journal of the American Medical Association* 1959, **169** 1286.

Friedman, M and Rosenman, R H (1974) *Type A Behaviour and Your Heart,* Knopf, New York.

Friedman, M, Rosenman, R H and Strauss, R 'The Relationship of Behaviour Pattern A to the State of the Coronary Vasculature', *American Journal of Medicine* 1968, **44** 525.

Fromm, E (1978) *To Have or to Be?* Jonathan Cape Ltd, London.

Gillon, R 'Aids and Medical Confidentiality', *British Medical Journal* 1987, **294** 1675–77.

Goodwin, P (1984) *Can You Avoid Cancer?* British Broadcasting Corporation, London.

Grant, M, Plant, M and Williams, A (Ed) (1983) *Economics and Alcohol: Consumption and Controls,* Croom Helm, London.

Gray, J A M and Fowler, G (1984) *Essentials of Preventive Medicine,* Blackwell Scientific Publications, Oxford.

Hagan, L W and Waterson, M J (1983) *The Impact of Advertising on the United Kingdom Alcohol Drink Market,* Advertising Association, London.

Hancock, G and Carim, E (1986) *Aids: The Deadly Epidemic,* Victor Gollancz Ltd, London.

Havighurst, C C, Helms, R B, Bladen, C and Pauly, M V (1988) *American Health Care: What are the Lessons for Britain?* The Institute of Economic Affairs, London.

Hayes, R and Watts, R. (1986) *Corporate Revolution,* Heinemann, London.

Health Education Authority (1987) *Smoking Policies at Work,* London.

Health Promotion Research Trust (1987) *The Health and Lifestyle Survey,* London.

Health Research Institute (1987) *Biennial Survey Results, Participants Report,* Walnut Creek, California.

Heather, N and Robertson, J (1986) *Problem Drinking: The New Approach,* Penguin Books, Harmondsworth.

Higgins, I T T, Mahan, C M and Wynder, E L 'Lung Cancer Among Cigar and Pipe Smokers', *Preventive Medicine* 1988, **17** 116–28.

Høstmark, A T, Bjerkedal, T, Kierulf, P, Flaten, H and Ulshagen, K 'Fish Oil and Plasma Fibrinogen', *British Medical Journal* 1988, **297** 180–81.

HMSO (1989) *Working for Patients*, London 1989.

Inglis, B (1983) *The Diseases of Civilisation*, Paladin, Granada Publishing Ltd, London.

Jennett, B (1986) *High Technology Medicine*, Oxford University Press, Oxford.

Kavanagh, T (1985) *The Healthy Heart Program*, Key Porter Books, Toronto.

Kennedy, I (1983) *The Unmasking of Medicine*, Paladin, Granada Publishing Ltd, London.

Keys, A (Ed) 'Coronary Heart Disease in Seven Countries', *Circulation* 1970, 41–2, Supplement 1: 186–95.

Kramsch, D M, Aspen, A J and Abramovitz, B M 'Reduction of Coronary Atherosclerosis by Moderate Conditioning Exercise in Monkeys on an Atherogenic Diet', *New England Journal of Medicine* 1981, **305** 1483–9

Kristein, M M 'Health Care Costs and Preventive Medicine', *Preventive Medicine* 1982, **11** 729–32.

Kristein, M M 'How Much Can Business Expect to Profit From Smoking Cessation?' *Preventive Medicine* 1983, **12** 358–81.

Kristein, M M 'The Economics of Health Promotion at the Worksite', *Health Education Quarterly* 1982, **9** Special Supplement: 27–36.

Kristenson, H, Ohlin, H, Hulten-Nosslin, M, Trell, E and Hood, B 'Identification and Intervention of Heavy Drinkers in Middle-aged Men: Results and Follow-up of 24–60 Months of Long-term Study with Randomized Controls', *Journal of Alcholism, Clinical and Experimental Research* 1983, 203–9.

Laing, E T (1983) *The Economics of Smoking and Health*, Health Education Council, London.

Larson, E B and Bruce R A 'Exercise and Aging', *Annals of Internal Medicine* 1986, **105** 783–5.

Leviton, L C 'The Yield from Work Site Cardiovascular Risk Reduction', *Journal of Occupational Medicine*, USA 1987, **29** 931–5.

Lipid Research Clinics Program 'The Lipid Research Clinics Coronary Primary Prevention Trial Results: 1 Reduction in Incidence of Coronary Heart Disease', *Journal of the American Medical Association* 1984, **251** 351–64.

McKeown, T (1979) *The Role of Medicine*, Basil Blackwell Ltd, Oxford.

McKeown, T and Lowe, C R (1974) *An Introduction to Social Medicine* (2nd Ed), Blackwell Scientific Publications, Oxford.

McNeill, A (1983) *Guidelines on Alcohol Education in the Workplace*, Institute of Alcohol Studies, London.

Melville, J (1984) *The Tranquillizer Trap*, Fontana Paperbacks, London.

Morris, J N 'Exercise, Health and Medicine', *British Medical Journal* 1983, **286** 1597–8.

Morris, J N, Adam, C, Chave, S P W, Sirey, C, Epstein, L and Sheehan, D J 'Vigorous Exercise in Leisure-time and the Incidence of Coronary Heart Disease', *Lancet* 1973, **i** 333–9.

Morris, J N, Heady, J A, Raffle, D A B, Roberts, C G and Parks, J W 'Coronary Heart Disease and Physical Activity of Work', *Lancet* 1953, **2** 1053–7.

National Advisory Committee on Nutrition Education (1983) *Proposals for Nutritional Guidelines for Health Education in Britain*, Health Education Council, London.

National Audit Office (1989) *National Health Service: Coronary Heart Disease*, HMSO, London.

Nelson, H (1987) *The Economic Consequences of Smoking in Northern Ireland*, Ulster Cancer Foundation, Ulster.

Office of Disease Prevention and Health Promotion (1986) *Worksite Health Promotion: A Bibliography of Selected Resources*, US Department of Health and Human Services, Washington DC.

Office of Disease Prevention and Health Promotion (1987) *National Survey of Worksite Health Promotion Activities: A Summary*, US Department of Health and Human Services, Washington DC.

Office of Disease Prevention and Health Promotion (1987) *'Staying Healthy': A Bibliography of Health Promotion Materials*, US Department of Health and Human Services, Washington DC.

Office of Health Economics (1983) *Keep on Taking the Tablets?*, Office of Health Economics, Paper No 21, London.

Office of Health Economics (1987) *Coronary Heart Disease: The Need for Action*, Office of Health Economics, Paper No 86, London.

Office of Health Economics (1988) *'Stroke'*, Office of Health Economics, Paper No 89, London.

O'Neill, P (1983) *'Health Crisis 2000'* Heinemann Medical Books Ltd, London.

Opatz, J P (Ed) (1987) *Health Promotion Evaluation: Measuring the Organisational Impact,* The National Wellness Institute, Stevens Point, Wisconsin.

Owen, D (1988) *Our NHS,* Pan Books Ltd, London.

Paffenbarger, R S and Hale, W E 'Work Activity and Coronary Heart Mortality', *New England Journal of Medicine* 1975, **292**(ii) 545–50

Paffenbarger, R S, Hyde, R T, Wing, A L and Hsieh, C C 'Physical Activity, All-Cause Mortality, and Longevity of College Alumni', *New England Journal of Medicine* 1986, **314** 605–13.

Paffenbarger, R S, Hyde, R T, Wing, A L and Steinmetz, C H 'A Natural History of Athleticism and Cardiovascular Health', *Journal of the American Medical Assocation* 1984, **252** 491–5.

Paffenbarger, R S, Wing, A L and Hyde, R T 'Physical Activity as an Index of Heart Attack Risk in College Alumni', *American Journal of Epidemiology* 1978, **108** 161–75.

Pells, S and Fayerweather, W 'Trends in the Incidence of Myocardial Infarction and in Associated Mortality and Morbidity in a Large Employed Population, 1957–83', *New England Journal of Medicine* 1985, **312** 1005–11.

Redwood, J (1988) *In Sickness and in Health,* Study No 95, Centre for Policy Studies, London.

Richman, J (1987) *Medicine and Health,* Longman Group (UK) Ltd, Harlow.

Roberts, J L and Graveling, P A (Ed) (1985) *The Big Kill. Smoking Epidemic in England and Wales,* North West Regional Health Authority for the Health Eduction Council and the British Medical Association.

Robertson, I and Heather N (1986) *'Let's Drink to Your Health',* The British Psychological Society, London.

Roebuck, E J 'Mammography and Screening for Breast Cancer', *British Medical Journal* 1986, **292** 223–6.

Rosenman, R H, Brand, R J, Jenkins, C D, Friedman, M, Strauss, K and Wurm, M 'Coronary Heart Disease in the Western Collaborative Group Study: Final Follow-up Experience of 8½ Years', *Journal of the American Medical Association* 1975, **233** 872–77.

Rosenman, R H, Friedman, M and Strauss, R 'A Predictive Study of Coronary Heart Disease: The Western Collaborative Group Study', *Journal of the American Medical Association* 1964, **189** 15–22.

Royal College of General Practitioners (1986) *Alcohol – A Balanced View,* Report from General Practice 24, London.

Royal College of Physicians (1971) *Smoking and Health Now. Second Report,* Pitman Press, London.

Royal College of Physicians (1983) *Health or Smoking?,* Pitman Press, London.

Royal College of Physicians (1987) *A Great and Growing Evil: The Medical Consequences of Alcohol Abuse,* Tavistock Publications, London.

Royal College of Psychiatrists (1986) *Alcohol – Our Favourite Drug,* Tavistock Publications, London.

Royal Society of Medicine *The AIDS Letter,* (Published bi-monthly) Royal Society of Medicine Services Ltd, London.

Schumacher, E F (1974) *'Small is Beautiful'* Abacus Edition, Sphere Books Ltd, London.

Shaper, A G (1988) *Coronary Heart Disease – Risks and Reasons,* Current Medical Literature Ltd, London.

Shaper, A G, Pocock, S J, Walker, M, Cohen, N M, Wale, C J and Thomson, A G 'British Regional Heart Study: Cardiovascular Risk-factors in Middle-aged Men in 24 Towns', *British Medical Journal* 1982, **283** 179–86.

Shekelle, R B, Gayle, M and Ostfeld, P O 'Hostility, Risk of Coronary Heart Disease and Mortality', *Psychosomatic Medicine* 1983, **45** 109–14.

Shekelle, R B, Hulley, S B and Neaton, J D. 'The MRFIT Behaviour Pattern Study II. Type A Behaviour and Incidence of Coronary Heart Disease', *American Journal of Epidemiology* 1985, **122** 559–70.

Singer, A and Szarewski, A (1988) *Cervical Smear Test: What Every Woman Should Know,* Macdonald Optima, London.

Skinner, H A, Holt, S, Sheu, W J and Israel Y 'Clinical versus Laboratory Detection of Alcohol Abuse: The Alcohol Clinical Index', *British Medical Journal* 1986; **292** 1703–8.

Sloan, R P, Gruman, J C and Allegrante, J P (1988) *Investing in Employee Health: A Guide to Effective Health Promotion in the Workplace,* Jossey-Bass, San Francisco.

Smith, R 'Alcohol: A New Report but Still Going Backwards', *British Medical Journal* 1986, **293** 971–2.

Smith, R (1987) *Unemployment and Health*, Oxford University Press, Oxford.

Stamler, J. 'Coronary Heart Disease: Doing the "Right Things"', *New England Journal of Medicine* 1985, **312** 1053–55.

Taylor, P (1985) *The Smoke Ring: Tobacco, Money and Multinational Politics*, Sphere Books, London.

Truswell, S A (1986) *ABC of Nutrition*, British Medical Association, London.

UK National Case-Control Study Group 'Oral Contraceptive Use and Breast Cancer Risk in Young Women', *Lancet* 1989, **i** 973–82.

US Surgeon General's Report (1985) *The Health Consequences of Smoking: Cancer and Chronic Lung Disease in the Workplace*, US Department of Health and Human Services, Washington DC.

Whitney, R (1988) *National Health Crisis*, Shepheard-Walwyn Ltd, London.

Williams, C J 'Preventing Lung Cancer', *British Medical Journal* 1987, **295** 1433–4.

INDEX